Essential Physics Workbook

by

Andrew Duffy
Department of Physics
Boston University

© 2019

QR code for a page that links to all the pre-session videos

ESSENTIAL PHYSICS

Volume I

Mechanics

Momentum and Energy

Rotation

Fluids

Thermodynamics

Gravitation

Table of Contents

Pre-session worksheets

Exploring Motion

Learning goals – by the end of this section you should be able to
- Draw and interpret position-vs.-time graphs
- Draw and interpret velocity-vs.-time graphs
- Be able to go back and forth between different representations of motion

An ultrasonic motion sensor works like a bat. A bat sends out high frequency sound waves that bounce off objects. The motion sensor does the same, allowing us to measure how far away an object is and how fast, and in what direction, it is moving.

Step 1 – Play: Spend a few minutes messing around with collecting data so you have some idea of how things work.

Step 2 – Individual predictions
On the graphs below, sketch what you think the position-vs.-time and velocity-vs.-time graphs will look like for the following motions.
- A - Walk slowly and steadily in the positive direction
- B - Walk quickly and steadily in the positive direction
- C- Walk slowly and steadily in the negative direction
- D - Walk quickly and steadily in the negative direction

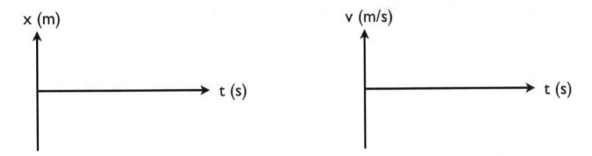

Step 3 – Group discussion
Compare your predictions with your group. It is important to **not be judgmental**, and to **listen respectfully** to your colleagues and **contribute your own ideas**. Everyone should be comfortable sharing. It's fine to be wrong – then you learn something.

Step 4 – If necessary, revise your predictions below

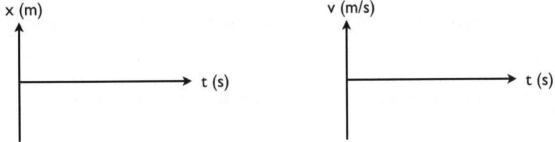

Step 5 – Carry out the measurements

On the graphs below, sketch the position-vs.-time and velocity-vs.-time graphs for the following motions. You don't have to sketch exactly what the computer shows. You can ignore glitches at the beginning and end, and **sketch what the graph would look like if the person was able to walk really smoothly**.

- A - Walk slowly and steadily in the positive direction
- B - Walk quickly and steadily in the positive direction
- C- Walk slowly and steadily in the negative direction
- D - Walk quickly and steadily in the negative direction

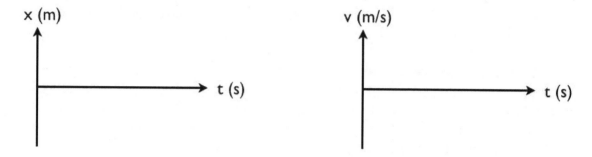

Step 6 – Thinking about the position-vs.-time graph

How does how fast the person walks show up on the position-vs.-time graph?

How does the steadiness of the motion show up on the position-vs.-time graph?

How does the direction of the motion show up on the position-vs.-time graph?

Step 7 – Thinking about the velocity-vs.-time graph

How does how fast the person walks show up on the velocity-vs.-time graph?

How does the steadiness of the motion show up on the velocity -vs.-time graph?

How does the direction of the motion show up on the velocity -vs.-time graph?

Follow-up: translating between the two graphs

Draw a position-vs.-time graph that corresponds to the following motions.

(a) The velocity-vs.-time graph is constant and positive.
(b) The velocity-vs.-time graph is constant and negative.
(c) The velocity-vs.-time graph is constant and zero.
(d) The velocity-vs.-time graph shows a large initial speed, but the speed decreases at a constant rate, eventually reaching zero.

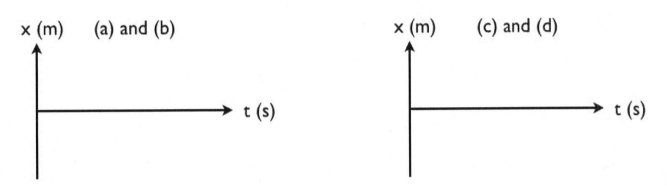

Draw a velocity-vs.-time graph that corresponds to the following motions.

(e) The position-vs.-time graph is constant and positive.
(f) The position-vs.-time graph is constant and negative.
(g) The position-vs.-time graph slopes upward, with a constant slope.
(h) The position-vs.-time graph slopes downward, with a constant slope.

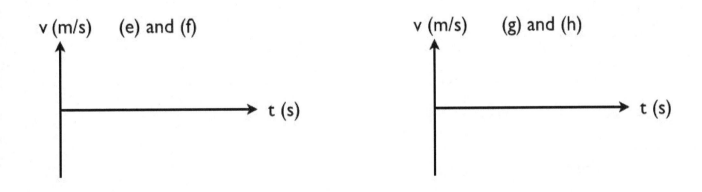

Pre-session: Vectors

Direct link: https://www.youtube.com/embed/uZXuojJRsoE

Define a vector:

Give examples of vectors:

Define a scalar:

Give examples of scalars:

Describe the tail-to-tip method (also called the tip-to-tail method):

True or false? The tail-to-tip method only works for two vectors. _____

Complete the table:

Vector	x-component	y-component
B		
C		
R = B + C		

Express R in unit-vector notation:

Find the magnitude of R:

Find the angle between R and the negative x-axis:

Vectors
Knowing how to work with vectors, add vectors, and distinguish between vectors and scalars will be an important part of the course.

What is a vector? Give a brief definition in words.

Give some examples of quantities that are vectors.

What is a scalar? Give a brief definition in words.

Give some examples of quantities that are scalars.

You have two vectors of length 4.0 m and 7.0 m (m stands for meters). You get to choose the direction of each vector. When you add these two vectors tip-to-tail, what is...

...the magnitude of the largest resultant vector you can obtain?

Your answer: _____   Your answer after discussion: _____

...the magnitude of the smallest resultant vector you can obtain?

Your answer: _____   Your answer after discussion: _____

Can you obtain resultant vectors of all possible lengths between the smallest and largest values? Explain.

Vectors and Vector Addition

Learning goals – by the end of this section you should be able to
- Add vectors using the tip-to-tail method
- Find vector components, and add vectors using components

You have three vectors of length 4.0 m, 7.0 m, and 9.0 m (m stands for meters). You get to choose the direction of each vector. When you add these three vectors tip-to-tail, what is…

…the magnitude of the largest resultant vector you can obtain?

  Your answer: _____   Your answer after discussion: _____

…the magnitude of the smallest resultant vector you can obtain?

  Your answer: _____   Your answer after discussion: _____

Think about this at home: You have three vectors of length 4.0 m, 7.0 m, and 13.0 m You get to choose the direction of each vector. When you add these three vectors tip-to-tail, what is…

…the magnitude of the largest resultant vector you can obtain?

…the magnitude of the smallest resultant vector you can obtain?

Consider the following right-angled triangle. The hypotenuse has a length of 5 m, and the angle opposite side a is 30°. What is the length of side a and side b?

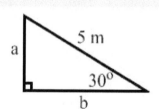

What is the relationship between a, b, and c in a general right-angled triangle?

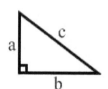

What is the relationship between a, b, and c in a triangle that has an angle θ_c opposite side c ? Show that the Pythagorean Theorem is a special case of this relationship for $\theta_c = 90°$.

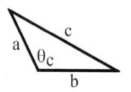

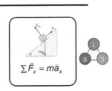

Vector Addition

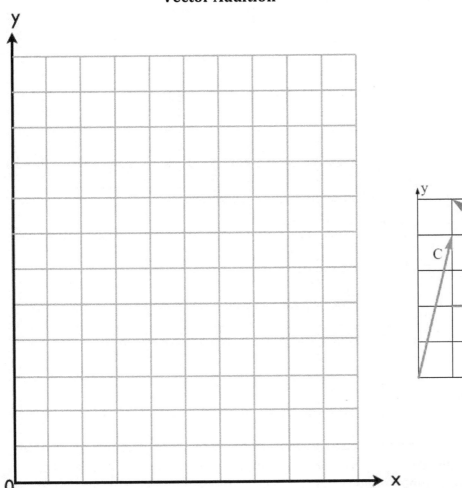

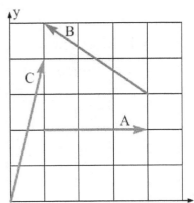

Draw the following on the grid:

1. A + C   2. 2A + B   3. C – B

4. Combine an integer number of A's and an integer number of C's so the resultant has no x-component

5. Combine an integer number of B's and an integer number of C's so the resultant has no y-component

Vector Components - *A standard way to add vectors is to use components.*

Let's say that the squares on the grid at right measure 1 m × 1 m.

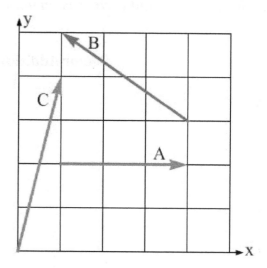

What are the *x* and *y* components of vector ***B*** ?

Vector	*x*-component	*y*-component
$\vec{B}$	$B_x =$	$B_y =$
$\vec{C}$	$C_x =$	$C_y =$
Resultant $\vec{R} = \vec{B} + \vec{C}$	$R_x =$	$R_y =$

Consider the resultant vector obtained by adding the vectors ***B*** and ***C***. Fill in the table above, which should help you find the components of the resultant vector.

What is the magnitude and direction of this resultant vector? (Feel free to write down an expression for the sine, cosine, or tangent of an appropriate angle if you can't determine an angle directly.)

Pre-session: Speed and Velocity

Direct link: https://www.youtube.com/embed/jftA-F0tbYE

Define speed:

Define velocity:

Equation for average speed:

Equation for average velocity:

Equation for instantaneous velocity:

How do you find the instantaneous speed if you have the instantaneous velocity?

How do you find instantaneous velocity from a position-vs.-time graph?

How do you find displacement from a velocity-vs.-time graph?

Velocity and Speed
What is the difference between average speed, average velocity, instantaneous speed and instantaneous velocity?

$$\text{average speed} = \frac{\text{total distance}}{\text{total time}} \qquad\qquad \text{average velocity} = \frac{\text{net displacement}}{\text{total time}}$$

On your way to class one morning, you leave home and walk at 3.0 m/s east towards campus. After exactly one minute, you realize that you've left your physics assignment at home, so you turn around and run, at 6.0 m/s, back to get it. You're running twice as fast as you walked, so it takes half as long (30 seconds) to get home again.

Note that you covered 180 m before turning around.

 Individual exercise - What is your average speed for the round trip?

 What is your average velocity for the round trip?

 Now, compare your answers with the other two members of your group.

Plot a graph of speed as a function of time and velocity as a function of time.

Understanding graphs

Learning goals – by the end of this section you should be able to
- Interpret graphs to calculate average and instantaneous values of speed and velocity
- Apply the constant-acceleration equations to analyze typical situations involving motion.
- Draw and interpret graphs (such as of velocity vs. time or acceleration vs. time) to help analyze motion situations.

Individual exercise - Consider the graph at right, showing your position as a function of time as you move along a straight sidewalk.

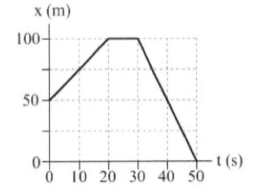

What is your velocity at t = 10 s? _____

What is your velocity at t = 25 s? _____

What is your displacement during the 10-second interval from t = 20 s to t = 30 s?

_____

What is your average velocity over the 50-second period? _____

What is your average speed over the 50-second period? _____

Now, compare your answers with the other two members of your group.

What is your velocity at t = 10 s? _____

What is your velocity at t = 25 s? _____

What is your displacement during the 10-second interval from t = 20 s to t = 30 s?

_____

What is your average velocity over the 50-second period? _____

What is your average speed over the 50-second period? _____

Pre-session: Acceleration

Direct link: https://www.youtube.com/embed/J5WcOCHmJJg

Define acceleration in words:

Equation for acceleration:

What happens when the acceleration and the velocity are in the same direction?

What happens when the acceleration and the velocity are in opposite directions?

What happens to the velocity when there is no acceleration?

Write out the constant-acceleration equations:

How do you find acceleration from a velocity-vs.-time graph?

How do you find the change in velocity from an acceleration-vs.-time graph?

Sketch a motion diagram for an object experiencing motion with constant acceleration, starting from rest.

Applying the constant-acceleration equations

$$x = x_i + v_i t + \frac{1}{2} a t^2 \qquad v = v_i + at \qquad v^2 = v_i^2 + 2a(x - x_i) \qquad \Delta x = v_{av} t$$

Now we'll get some practice using the equations. Be methodical. In each case begin by:
1. Drawing a sketch of the situation.
2. Choosing an origin and a positive direction and marking them on the sketch.
3. Setting up a data table with everything you know.
4. Only then figuring out which of the constant-acceleration equations to apply.

EXAMPLE 1: A cyclist has an initial velocity of 4.0 m/s directed south. The cyclist then accelerates at 2.0 m/s² south for 3.0 seconds.

(a) What is the cyclist's velocity at the end of the 3.0-second acceleration period? _____

(b) How far does the cyclist travel during the 3.0-second acceleration period? _____

First, complete the four steps listed above, and then answer parts (a) and (b).

Compare your answers with those of the other members of the group. See how many different ways you can come up with to solve the problem.

$$x = x_i + v_i t + \frac{1}{2}at^2 \qquad v = v_i + at \qquad v^2 = v_i^2 + 2a(x - x_i) \qquad \Delta x = v_{av}t$$

Begin by:
1. Drawing a sketch of the situation.
2. Choosing an origin and a positive direction and marking them on the sketch.
3. Setting up a data table with everything you know.
4. Only then figuring out which of the constant-acceleration equations to apply

EXAMPLE 2: You are driving your car at 20 m/s when you see a deer in the road 60 m ahead. It takes you 1.0 seconds before you apply the brakes, but then the car slows down and comes to a stop. Assuming the car's acceleration is constant, what magnitude acceleration (at least) is required to avoid hitting the deer? _____

First, complete the four steps listed above, and then answer the question.

Compare your answer with those of the other members of the group. See how many different ways you can come up with to solve the problem.

A graphing exercise for motion with constant acceleration

 Consider the motion of an object (a ball, or a set of keys, for example) that falls straight down after you drop it from rest. Plot a velocity vs. time graph for it.

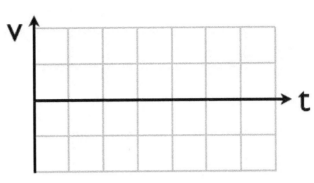

 Now, compare your graph with the graphs drawn by the other two members of your group.

Show the group consensus graph here.

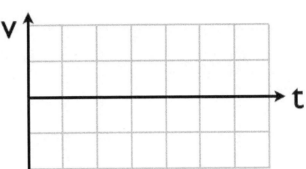

Do you see any similarity between this velocity vs. time graph and one of the graphs that corresponds to constant-velocity motion? (for instance, are the shapes similar?)

 Acceleration is related to velocity in the same way that velocity is related to position. Based on this statement, and what we know about velocity, give a mathematical definition of acceleration and plot the acceleration vs. time graph for the dropped object.

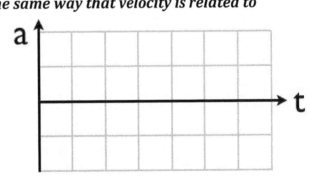

Now, compare your graph with the graphs drawn by the other two members of your group.

Show the group consensus graph here.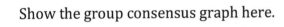

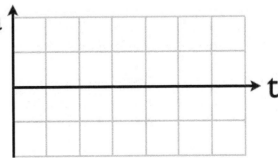

Now consider a different motion, the motion of a ball that is thrown straight up into the air, and caught again on the way down. Plot a velocity vs. time graph for it. Then, plot an acceleration vs. time graph for it.

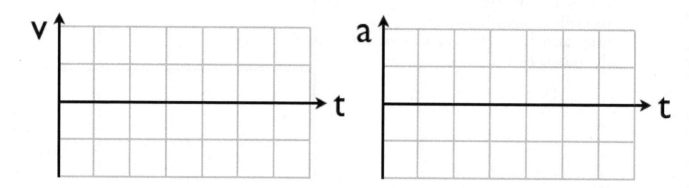

Now, compare your graph with the graphs drawn by the other two members of your group. Show the group consensus graph below.

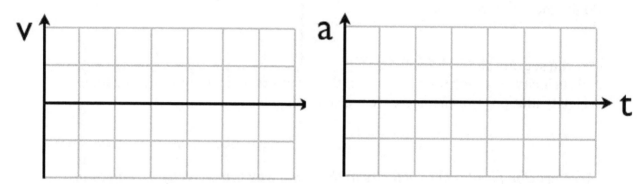

Can you tell from either graph when the object reaches its highest point? If so, how?

The object comes to rest for an instant at its highest point. What, if anything, does the fact that the ball comes to rest at that point tell us about its acceleration at that point?

Which of these motions would have similar graphs (in terms of shape, not necessarily numerical values)?

[ ] A toy car rolls up a ramp and then back down (compare to this page)
[ ] You accelerate your car at a constant rate from rest (compare to previous page)
[ ] A ball rolls down a ramp (compare to previous page)

Sample test question

Two trains, a freight train and a commuter train, are traveling in the same direction on parallel tracks. At $t = 0$ s, the front of the freight train is 56.0 m **behind** the front of the commuter train. Be consistent, and measure everything from the front of each train. The graph above shows the velocities of the two trains from $t = 0$ s to $t = 60$ s. Note that the commuter train has a constant acceleration until $t = 24.0$ s, and then travels at constant velocity.

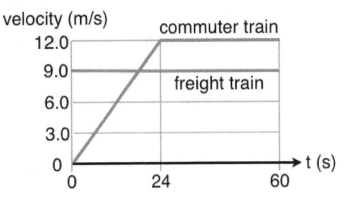

[3 points] (a) At some instant during the 60.0 s period, the freight train passes the commuter train. What time does that happen, the first time the two trains have the same position? (*Hint: start by writing out position equations for each train.*)

[3 points] (b) The freight train then moves ahead of the commuter train. What is the maximum distance the freight train gets ahead of the commuter train? (*Hint: when does this occur?*)

[3 points] (c) The commuter train eventually catches and passes the freight train. What time does that happen, the second time the two trains have the same position?

Pre-session: Force

Direct link: https://www.youtube.com/embed/wmIQ-ZIos-k

Define force:

A force is a   [ ] vector   [ ] scalar

Forces are associated with …

Forces come in …

The unit we use for force is the _____ .

What is this unit in terms of kg, m, and s?

A good example of a force that does not require contact between objects is …

What is the name for g on Earth and how big is it at the Earth's surface?

The force of tension goes _____ along the string or rope.

The normal force is the component of the contact force that is _____
to the contact interface.

Objects lose contact with one another when the normal force goes to _____.

True or false? If an object has a weight of 15 N, the normal force acting on it must
also be 15 N.

                    [ ] True   [ ] False

Forces and Free-Body Diagrams
What is a force? How do you draw a free-body diagram?

Learning goals – by the end of this section you should be able to
- Define force
- Draw a free-body diagram, and identify all the forces on it
- Identify Newton's third law force pairs and relate them to Newton's third law.

Define **force**.

Sketch free-body diagrams for two objects that are falling straight down to the ground.

A **free-body diagram** is a diagram that shows all the forces that are applied to a particular object or a particular system. Each force on a free-body diagram is represented by an arrow pointing in the direction of the force. The length of an arrow should be proportional to the magnitude of the force. We can also say that **a force comes from an interaction** – account for all the interactions the object is experiencing as it falls.

The first object has a mass *m*.	The second object has a mass 3*m*.

If they are released simultaneously from the same height above the ground, which object hits the ground first? Why?

Check your free-body diagrams against those of the other students in your group.

What (if anything) is the same for these two objects?

This is a good time to bring up Newton's second law. Write out the equation for Newton's second law and use it to find the acceleration of each of the objects above.

The lab scenario

Two carts are hooked together on a horizontal track. You exert a horizontal force F to the right on a red cart, and the red cart pulls on the blue cart, so both carts accelerate together to the right. Neglect friction.

The red cart has a mass m, the blue one has a mass $3m$.

Sketch your own free-body diagrams:

The two-cart system	Blue cart	Red cart

Consensus free-body diagrams:

The two-cart system	Blue cart	Red cart

Calculate the acceleration, in terms of F and m, and determine the magnitude of each horizontal force, in terms of F.

Four FBD's. Sketch a free-body diagram for each of the following situations. Check your free-body diagrams with the other students in your group.

1. An object is at rest in outer space, billions of kilometers from anything.	2. An object is drifting at constant velocity to the right in outer space, billions of kilometers from anything.
3. A box remains at rest on a horizontal table.	4. A hockey puck slides at constant velocity to the right on a frictionless ice rink.

Your two diagrams on the left should be consistent with one another, because they both relate to objects at rest. In what way are they consistent?

Your two diagrams on the right should also be consistent with one another. In what way are they consistent?

Newton's second law allows us to be quantitative. Newton's first law really tells us the same thing in a qualitative way. What is Newton's first law?

When you are on an airplane, when is it that you feel yourself pressed back in your seat? When you're standing inside a bus or train, when do you need to hold on tight?

As you draw a free-body diagram ask yourself these questions:
1. For each force I draw on the free-body diagram, what applies the force?
2. Considering the object's motion, is the overall free-body diagram consistent with Newton's laws?

Are your four free-body diagrams above consistent with Newton's laws of motion? Correct your diagrams, if necessary, above.

Pre-session: Free-body diagrams

Direct link (part 1): https://www.youtube.com/embed/_IUdM3hw5fs

Two questions (rules of thumb) you should ask yourself when drawing free-body diagrams are:

(1)

(2)

A small box sits on a larger box. You support the two boxes by having the larger box rest on your arms. Draw three free-body diagrams:

Small box (on top)	Large box (underneath)	The two-box system

On the large box's free-body diagram, how should you represent the force of the small box acting on the large box?

[ ] As *mg*, the force of gravity acting on the small box
[ ] As a normal force   [ ] Either of these – they're equivalent

Direct link (part 2): https://www.youtube.com/embed/zZkhF8q7cuM

Again draw three free-body diagrams, this time for the boxes accelerating upward. Circle any forces that have changed from the original diagrams above.

Small box (on top)	Large box (underneath)	The two-box system

Write out Newton's third law:

A tug-of war

Mark (on the left) and Adam (on the right) are having a tug-of-war. Mark is winning – assume that he is accelerating to the left. How is it possible to win a tug-of-war, given what we learned previously?

Sketch two free-body diagrams, one for Mark and one for Adam, showing all the forces acting on each.

Identify forces that are equal-and-opposite because of Newton's third law.

Identify forces that are equal-and-opposite because of Newton's second law.

Apply Newton's second law to the Mark free-body diagram and to the Adam free-body diagram.

Which force(s) are critical to determining who wins the tug-of-war?

More on Forces and Free-Body Diagrams

Learning goals – by the end of this section you should be able to
- Draw, interpret, and explain free-body diagrams for individual objects and for a system of more than one object.
- Construct one or more force equations from one or more free-body diagrams, and solve for unknown parameters.

Elevator Physics
Let's get some more practice drawing free-body diagrams.

Consider the following situation. You are in an elevator. The elevator and you are **at rest**. For this situation sketch three free-body diagrams.

A free-body diagram for you.	A free-body diagram for the elevator.	A free-body diagram for the system consisting of you plus the elevator.

Check your free-body diagrams against those of the other students in your group.

Consider the following situation. You are in an elevator. The elevator and you are **moving up with constant velocity**. For this situation sketch three free-body diagrams.

A free-body diagram for you.	A free-body diagram for the elevator.	A free-body diagram for the system consisting of you plus the elevator.

Circle any differences between this set of FBD's and your previous set.

Check your free-body diagrams against those of the other students in your group.

Consider the following situation. You are in an elevator. The elevator and you have an acceleration that is directed up. For this situation, sketch three free-body diagrams.

A free-body diagram for you.	A free-body diagram for the elevator.	A free-body diagram for the system consisting of you plus the elevator.

Circle any differences between this set of FBD's and your first set.

Check your free-body diagrams against those of the other students in your group. Re-draw them here if you need to.

A free-body diagram for you.	A free-body diagram for the elevator.	A free-body diagram for the system consisting of you plus the elevator.

Two Boxes

Two boxes are initially at rest, and side-by-side, on a table. The green box, on the left, has a mass of 3.0 kg. The blue box, on the right, has a mass of 2.0 kg. A hand then applies a constant force of 15.0 N to the right on the green box, making both boxes accelerate to the right. Assume there is no friction. Use $g = 10$ N/kg.

For this situation sketch three free-body diagrams.

Check your free-body diagrams against those of the other students in your group.

A free-body diagram for the green box.	A free-body diagram for the blue box.	A free-body diagram for the system consisting of the two boxes together.

Use your free-body diagrams, and Newton's second law, to determine the following:

    (a) The acceleration of the blue box.

    (b) The force the green box applies to the blue box.

    (c) The force the blue box applies to the green box.

Pre-session: The independence of X and Y

Direct link: https://www.youtube.com/embed/U_smADN9vII

A motion diagram is a diagram with dots recording the position of an object at regular time intervals. Sketch two motion diagrams below, for

(1) a ball dropped from rest, which falls straight down under the influence of gravity, and (2) a ball launch, with an initial horizontal velocity, from the same point as the first ball.

Explain which ball hits the ground first, and why.

The Independence of X and Y

Learning goals – by the end of this section you should be able to
- Define what is meant by the independence of X and Y, and use this concept to compare motions.

Constant-acceleration equations

Equation from 1-D	2-D: X equations	2-D: Y equations
$v = v_i + at$	$v_x = v_{ix} + a_x t$	$v_y = v_{iy} + a_y t$
$x = x_i + v_i t + \dfrac{1}{2}at^2$	$x = x_i + v_{ix} t + \dfrac{1}{2}a_x t^2$	$y = y_i + v_{iy} t + \dfrac{1}{2}a_y t^2$
$v^2 = v_i^2 + 2a(\Delta x)$	$v_x^2 = v_{ix}^2 + 2a_x(\Delta x)$	$v_y^2 = v_{iy}^2 + 2a_y(\Delta y)$

Motion with constant velocity: Let's look at the motion of an object that travels at a constant velocity of 20 m/s in the positive *x*-direction, starting from the origin. What is this object's acceleration?

On the x-axis below, plot the position of the object at 1-second intervals for six seconds. The boxes on the grid measure 5 m x 5 m. Start the object at the origin, at the bottom left, and go from there.

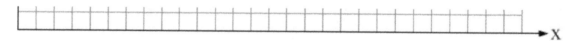

Motion with constant acceleration: Now, let's fire a ball straight up from the ground, giving it an initial velocity of 30 m/s up. Use *g* = 10 m/s² to keep the calculations simple.

What is the ball's acceleration?

How long does the ball spend in the air before it returns to the ground?

On the y-axis to the left below, plot the position of the ball at 1-second intervals, starting at the instant it leaves the ground until it returns to the ground. The boxes on the grid measure 5 m x 5 m. Start from the origin – at the bottom left. You may want to make a table of time, in 1-second intervals, and vertical position.

The squares on this grid measure 5 m x 5 m. Pay attention to the axes.

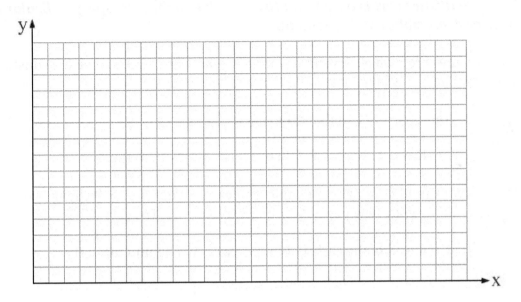

Re-create the two graphs from the previous page here.

Plot the position, at 1 second intervals, of a ball we fire from the origin so that its initial velocity has components of $\overrightarrow{v_{ix}}$ = 20 m/s in the +x direction and $\overrightarrow{v_{iy}}$ = 30 m/s straight up.

Which ball is in the air longer, the one we fired straight up or this one, that has both x and y components? Briefly explain your answer.

Now plot the position, at 1 second intervals, of a ball we fire from the origin so that its initial velocity has components of $\overrightarrow{v_{ix}}$ = 10 m/s in the +x direction and $\overrightarrow{v_{iy}}$ = 30 m/s straight up. Use a different symbol to mark these points on the diagram.

A Ballistics Cart
A ballistics cart is a cart that fires a ball in a direction perpendicular to its base and, possibly, catches it again.

Case 1 – the cart is on a horizontal track. The cart is pushed, to give it a constant velocity. The ball is launched while the cart is moving.

Predict where you think the ball will land:

[ ] ahead of the cart   [ ] in the cart   [ ] behind the cart

Discuss your prediction with the other students in your group. If change your mind, state your new prediction, and explain why. The cart will land:

[ ] ahead of the cart   [ ] in the cart   [ ] behind the cart

Case 2 – the cart is on an incline. The cart is released from rest. The ball is launched while the cart is moving.

Predict where you think the ball will land:

[ ] ahead of the cart   [ ] in the cart   [ ] behind the cart

Discuss your prediction with the other students in your group. If change your mind, state your new prediction, and explain why. The cart will land:

[ ] ahead of the cart   [ ] in the cart   [ ] behind the cart

Case 3 – the cart is on an incline. The cart is given an initial velocity directed up the slope. The ball is launched while the cart is moving up the slope.

Predict where you think the ball will land:

[ ] ahead of the cart   [ ] in the cart   [ ] behind the cart

Discuss your prediction with the other students in your group. If change your mind, state your new prediction, and explain why. The cart will land:

[ ] ahead of the cart   [ ] in the cart   [ ] behind the cart

Pre-session: Projectile motion

Direct link: https://www.youtube.com/embed/X2C6MSEbUZU

What is the general equation for the maximum height reached by a projectile?

Which of these parameters determines the maximum height reached? Select all that apply.

[ ] The horizontal component of the initial velocity
[ ] The vertical component of the initial velocity
[ ] The value of g.

The following apply only when the projectile lands at the same height it was launched from. Under this condition...

... what is the equation for the time-of-flight of the projectile?

... what are two forms of the equation for the range of the projectile?

Define **range**.

Motion in Two Dimensions / Projectile Motion

Learning goals – by the end of this section you should be able to
- Analyze basic projectile motion situations.

Constant-acceleration equations

Equation from 1-D	2-D: X equations	2-D: Y equations
$v = v_i + at$	$v_x = v_{ix} + a_x t$	$v_y = v_{iy} + a_y t$
$x = x_i + v_i t + \frac{1}{2}at^2$	$x = x_i + v_{ix} t + \frac{1}{2}a_x t^2$	$y = y_i + v_{iy} t + \frac{1}{2}a_y t^2$
$v^2 = v_i^2 + 2a(\Delta x)$	$v_x^2 = v_{ix}^2 + 2a_x(\Delta x)$	$v_y^2 = v_{iy}^2 + 2a_y(\Delta y)$

EXAMPLE PROBLEM: 3.00 seconds after being launched from ground level with an initial speed of 25.0 m/s, an arrow passes just above the top of a tall tree. The base of the tree is 45.0 m from the launch point. Neglect air resistance and assume that the arrow lands at the same level from which it was launched. Use g = 10.0 m/s².

Begin by sketching a diagram of the situation, and then creating a table that separates the x-direction (horizontal) information from the y-direction information

	x-direction	y-direction
Positive direction		
Initial position	$x_i =$	$y_i =$
Initial velocity	$v_{ix} =$	$v_{iy} =$
Acceleration	$a_x =$	$a_y =$
Displacement and time		

(a) At what angle, measured from the horizontal, was the arrow launched? Feel free to find the sine, cosine, or tangent of the angle instead of the angle itself if you find that to be easier.

(b) How tall is the tree?

(c) How far from the launch point does the arrow land?

Whole vectors

The squares on this grid measure 5 m x 5 m.

First, plot the location, at 1-second intervals, of a ball launched from the origin. Assume gravity has been magically turned off, so the ball's acceleration is zero. The initial velocity has components of $\vec{v}_{ix}$ = 20 m/s in the +x direction and $\vec{v}_{iy}$ = 30 m/s straight up. The dots you plot represent the position of the tip of the vector $\vec{v}_i t$.

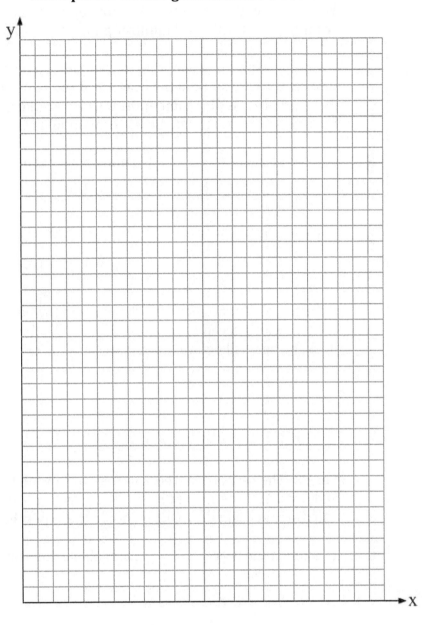

Now account for the acceleration due to gravity. At each of the six points you plotted (not including the origin) add a vector representing $\frac{1}{2}\vec{a}t^2$. Put a dot at the tip of each of these arrows – these dots represent the actual trajectory followed by the ball.

Hint: for the point at t = 1 s, accounting for gravity drops the point one grid unit (5 m) below the point without gravity.

We're using:

$$\vec{r} = \vec{r}_i + \vec{v}_i t + \frac{1}{2}\vec{a}t^2$$

Note that, instead of using components, we're really using whole vectors – keeping the vectors whole, and just adding them tail-to-tip.

Monkey and hunter
A classic physics demonstration

We'll do a gentler version of the monkey + hunter. Instead, we'll do a bear and a wildlife expert with a tranquilizer gun.

A wildlife expert is attempting to tranquilize a bear in a tree. The expert climbs a different tree so she and the bear are at the same vertical level. The expert takes aim, and fires. At the moment the dart leaves the gun, the bear lets go of the tree branch and drops straight down. How should the expert have aimed to hit the bear?

[ ] aim at the bear   [ ] aim above the bear   [ ] aim below the bear

[ ] aim at the bear   [ ] aim above the bear   [ ] aim below the bear

If the initial distance between the gun and the bear is L, and the dart comes out of the gun with a speed V, how long does it take the dart to reach the bear, and how far does the bear fall before the dart reaches the bear?

Case 2: The bear is in a tree, and the expert is on the ground. The expert takes aim, and fires. At the moment the dart leaves the gun, the bear lets go of the tree branch and drops straight down. How should the expert have aimed to hit the bear?

[ ] aim at the bear   [ ] aim above the bear   [ ] aim below the bear

[ ] aim at the bear   [ ] aim above the bear   [ ] aim below the bear

Explain your answer:

Consider the equation $v = v_i + at$. If you double the initial velocity for a ball that follows a parabolic path through the air, the time to reach maximum height...

[ ] stays the same   [ ] doubles   [ ] quadruples

Maximum height: What about the value of the maximum height? Prove what happens to the maximum height when the speed is doubled by:

  (a) using the equation that relates speed to height
  (b) using the equation that involves average velocity
  (c) drawing a velocity versus time graph, comparing the two motions
  (d) drawing a position versus time graph, comparing the two motions
  (e) sketching a motion diagram, comparing the two motions

(a)   (b)

(c)

(d)/(e)

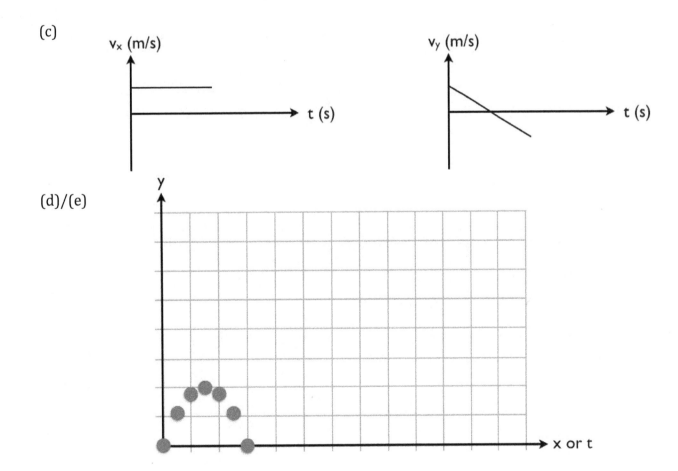

Pre-session: Friction

Direct link: https://www.youtube.com/embed/B6oEOZ9ny-g

Fill in the blank: Friction tends to oppose _____ motion between objects.

Kinetic friction

Equation:

Sketch the free-body diagram for a box sliding to the right on a horizontal table, slowing down because of kinetic friction.

Static friction

Equation:

Sketch the free-body diagram for a box at rest on a horizontal table, even though you exert a horizontal force on it to the right.

In general, what is the best answer to the question "What is the magnitude of the static friction force equal to in the case above?" ?

[ ] The coefficient of friction multiplied by the normal force.
[ ] The magnitude of the horizontal force you apply.

Why can static friction be harder to deal with than kinetic friction?

Friction

Learning goals – by the end of this section you should be able to
- Distinguish between kinetic friction and static friction.
- Incorporate friction in a force analysis.
- Analyze forces and motion in a situation of a block on a ramp

A kinetic friction example

You slide your textbook across a horizontal table. The book has an initial velocity of 1.0 m/s, directed to the right, and it comes to rest after sliding through a distance of 1.0 m. Sketch a free-body diagram of the book. Check your free-body diagram with the other members of your group. If necessary, show an amended free-body diagram here.

Write out Newton's second law for the vertical direction, and solve for the normal force (*i.e.*, get an expression for the normal force in terms of variables).

Write out Newton's second law for the horizontal direction, and then get an expression for the acceleration. Use your expression for the normal force to simplify your expression.

Using one of the constant-acceleration equations, find the numerical value of the acceleration.

Determine the numerical value of the coefficient of kinetic friction.

A block on an incline

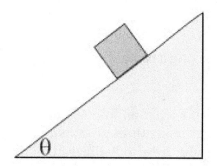

You place a block on a frictionless incline. With the block at rest, you let go of the block. What will the block do? Why?

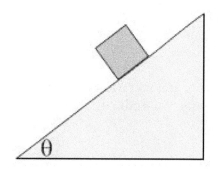

  Sketch a free-body diagram of the block.

 Check your free-body diagram with the other members of your group. If necessary, show an amended free-body diagram here.

What is a good coordinate system to use in this case? Why?

Is there only one coordinate system you can use in this case? Explain why or why not.

Using the recommended coordinate system, we will draw a new free-body diagram for the block, showing force components. All components should be parallel to one or the other of the coordinate axes.

Step 1 - Splitting a force into components should involve a right-angled triangle. Which force do we need to split into components here?

Show the relevant right-angled triangle on the diagram. The angle θ should appear in that triangle – show clearly how you know which angle is θ.

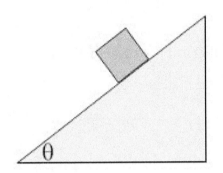

Step 2 - Now, draw a revised free-body diagram. Replace any force that is not parallel to a coordinate axis with its components, which should be parallel to the coordinate axes.

 Apply Newton's second law in the x-direction. Use the resulting equation to find an expression for the block's acceleration.

 Check this with the other members of your group.

 Apply Newton's second law in the y-direction. Use the resulting equation to find an expression for the normal force acting on the block.

 Check this with the other members of your group.

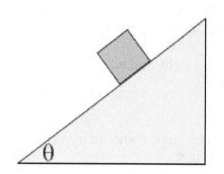

 Now, you place a box on a different ramp, and you observe that the box remains at rest after you let go of the box. Starting from the last free-body diagram on the previous page, show how the free-body diagram is modified to fit this particular situation.

 Check your free-body diagram with the other members of your group. If necessary, show an amended free-body diagram here.

 If you apply Newton's second law in the y-direction, you get exactly the same result for the normal force we got on the previous page. What is that result, and why is it the same expression?

 Apply Newton's second law in the x-direction. Use the resulting equation to find an expression for the force of friction that acts on the box.

 Check this with the other members of your group. Also, discuss whether or not the expression $F_S = \mu_S F_N$ applies in this case.

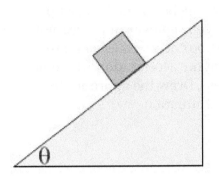

 Now, you give the box a small push down the ramp. After you remove your force, you observe that the box continues to accelerate down the ramp, but with an acceleration less than it would have if the ramp was frictionless. Draw the appropriate free-body diagram for this situation.

 Check your free-body diagram with the other members of your group. If necessary, show an amended free-body diagram here.

 If you apply Newton's second law in the y-direction, you get exactly the same result for the normal force we got previously. What is that result, and why is it the same expression?

 Apply Newton's second law in the x-direction. Use the resulting equation to find an expression for the acceleration of the box. Assume the coefficient of kinetic friction between the box and the surface is μ_K .

Check this with the other members of your group.

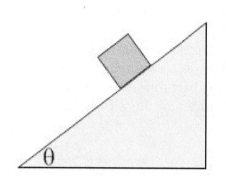

 Instead, you give the box a small push up the ramp. After you remove your force, you observe that the box accelerates at a rate that is different from what the acceleration would be if the ramp was frictionless. Draw the appropriate free-body diagram for this situation.

 Check your free-body diagram with the other members of your group. If necessary, show an amended free-body diagram here.

 Again, we get the same result for the normal force we got previously. What is that result?

 Apply Newton's second law in the x-direction. Use the resulting equation to find an expression for the acceleration of the box. Assume the coefficient of kinetic friction between the box and the surface is μ_K .

 Check this with the other members of your group.

Pre-session: A box on a ramp

Direct link: https://www.youtube.com/embed/Zlqq3leOwpk

A box is at rest on an incline, but if the angle were increased even a tiny bit, the box would start to slide.

Sketch the free-body diagram for the box.

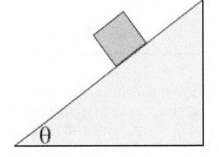

Using a coordinate system aligned with the ramp, show the free-body diagram with all force components aligned with one of the coordinate axes.

Write out Newton's second law for the direction perpendicular to the ramp.

Write out Newton's second law for the direction parallel to the ramp.

Put your equations together to get a simple expression for the coefficient of static friction.

Learning goals – by the end of this section you should be able to
- Be much more confident applying Newton's laws in various situations.

A box on a box

A large box of mass M sits on a horizontal surface.
A small box of mass m sits on top of the large box.
The coefficients of friction between all surfaces in
contact are:

$\mu_k = 0.40$ and $\mu_s = 0.50$

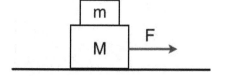

 A string of negligible mass is tied to the large box and a horizontal force F is applied to the string so that the two boxes accelerate to the right, with the small box maintaining its position on top of the large box at all times.

Draw three free-body diagrams for this situation. Label all arrows appropriately.

FBD for two-box system	FBD for the small box	FBD for the large box

 Check your free-body diagrams with the other members of your group. If necessary, show amended free-body diagrams here.

FBD for two-box system	FBD for the small box	FBD for the large box

Let's say that $M = 2.0$ kg and $m = 1.0$ kg, and we'll use the approximation that $g = 10$ N/kg to simplify the calculations.

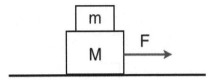

The coefficients of friction between all surfaces in contact are:
$\mu_k = 0.40$ and $\mu_s = 0.50$

If the force applied to the string is $F = 21$ N then the boxes accelerate together to the right with the small box maintaining its position on top of the large box. We want to solve for the acceleration of the system.

To find the acceleration of the system, which free-body diagram should we use?

[ ] two-box system   [ ] small box   [ ] large box

What is the acceleration of the system? Show all your work.

Under the conditions specified above, what is the magnitude of the force of friction acting on the small box? Show all your work.

What is the maximum value F can be without the small box sliding on the large box? Show all your work.

Same, or Different?

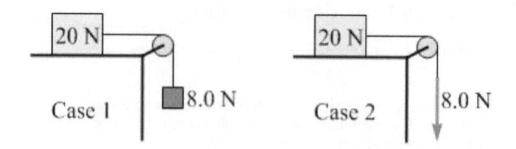

Case 1   8.0 N

Case 2   8.0 N

A 20 N block is on a frictionless surface. In case 1, the block is attached to an 8 N block by a string passing over a pulley (neglect friction, and neglect the pulley's mass). In case 2, you just pull on the string with an 8 N force.

In which case does the 20 N block have a larger acceleration?

[ ] Case 1   [ ] Case 2   [ ] Equal in both cases

In which case does the 20 N block have a larger acceleration?

[ ] Case 1   [ ] Case 2   [ ] Equal in both cases

In which case is the tension in the string larger?

[ ] Case 1   [ ] Case 2   [ ] Equal in both cases

In which case is the tension in the string larger?

[ ] Case 1   [ ] Case 2   [ ] Equal in both cases

Using g = 10 N/kg, solve for the acceleration and the tension in each case.

Pre-session: A box on a box on a ramp

No video for this one

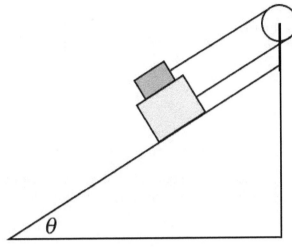

The green box is moving down the slope

The two boxes are connected by a string that goes around the pulley at the top of the ramp. If the green box (the bottom box) is moving down the slope, what does that mean the red box (the top box) must be doing?

[ ] moving up the slope   [ ] moving down the slope

[ ] at rest   [ ] could be any of these

Sketch the full free-body diagram for the top box.

Sketch the full free-body diagram for the bottom box.

Atwood's machine
How do we handle situations involving pulleys?

An Atwood's machine is a device that has two objects, one with mass M and one with mass m, connected by a string that passes over a pulley. Assume $M > m$, and that the pulley is massless and frictionless.

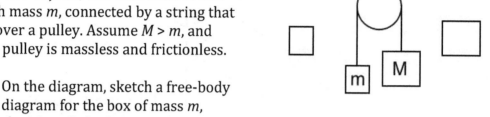

 On the diagram, sketch a free-body diagram for the box of mass m, showing all the forces acting on it.

For the box of mass m, show which direction you're taking to be positive. Apply Newton's second law to obtain a relationship between m, a (the acceleration), and the forces acting on the box.

On the diagram, sketch a free-body diagram for the box of mass M, showing all the forces acting on it.

For the box of mass M, show which direction you're taking to be positive. Apply Newton's second law to obtain a relationship between M, a (the acceleration), and the forces acting on the box.

Put the two equations together to get an expression for a, the acceleration, in terms of M, m, and g. Does your equation make sense? Test it and see.

Check everything with the other members of your group.

Pre-session: 10 boxes

No video for this one

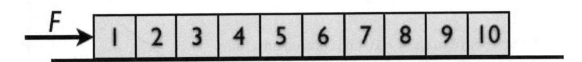

10 identical boxes, each of mass m, are **sliding to the right** on a horizontal surface. A force F, directed right, is applied to box 1.

First, analyze the situation assuming there is **no friction** between the boxes and the surface.

Find an expression for the acceleration of the system, in terms of F and m.

Find an expression for the force box 6 applies to block 7, in terms of F.

Now, analyze the situation assuming there **is friction** between the boxes and the surface, characterized by a kinetic coefficient of friction μ_K.

Find an expression for the acceleration of the system, in terms of F, m, g, and μ_K.

Find an expression for the force box 6 applies to block 7, in terms of F.

Inclined Atwood's machine

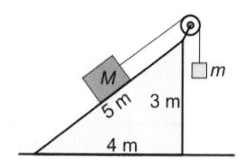

For this problem let's assume g = 10 N/kg.

A red block with a mass M = 10 kg is placed on a ramp that has the shape of the 3-4-5 triangle, with a height of 3.0 m and a width of 4.0 m. The red block is connected to a green box of mass m by a string that passes over a pulley at the top of the incline. The green box hangs vertically from the string.

Assume for the moment that there is **no friction** between the red block and the ramp. **Choose down the ramp to be positive for the red block.**

Our goal is to answer the following question - What value of m is required if we want the red block to remain at rest on the incline?

①   We'll start by drawing free-body diagrams.

FBD for the large block	FBD for the small box

If down the ramp is a positive direction for the large block, what direction should we take to be positive for the small box? Why?

Check your free-body diagrams with the other members of your group. Show your final free-body diagrams here.

FBD for the large block	FBD for the small box

Apply Newton's second law for the small box, to come up with a force equation.

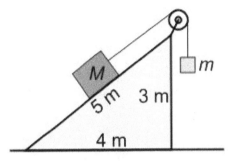

Apply Newton's second law twice for the large block, to come up with two force equations. Choose down the ramp as the positive *x* direction.

Put your equations together to solve for *m*, the mass of the small box that is needed to keep the system at rest.

Now, we introduce friction.
The coefficients are μ_s = 0.25 and μ_k = 0.5.

Having included friction, we'll need to know the normal force acting on the red block. Calculate the normal force.

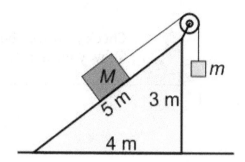

With no friction, m = 6 kg is sufficient to keep the system at rest. We'll keep m = 6 kg, but now that there is friction what is the friction force acting on the red block?

After completing the next page, return here to solve for the acceleration of the system when m = 1 kg.

Now, solve for the maximum value *m* can be without the system moving.

FBD for the large block	FBD for the small box

Solve for the maximum value *m* can be without the system moving.

Now, solve for the minimum value *m* can be without the system moving.

FBD for the large block	FBD for the small box

Solve for the minimum value *m* can be without the system moving.

Pre-session: Uniform circular motion

Direct link: https://www.youtube.com/embed/sJMBnE3XtYE

When an object travels in a circle at constant speed, does it accelerate?

[ ] Yes   [ ] No

Explain:

Centripetal acceleration

Magnitude:

Direction:

What is the relationship between the linear speed v and the angular speed ω?

Use that relationship to write out the centripetal acceleration equation in terms of angular speed instead of linear speed.

Uniform Circular Motion
Let's extend our application of Newton's laws to circular motion.

Learning goals – by the end of this section you should be able to
- Sketch free-body diagrams for circular motion situations.
- Apply Newton's second law in circular motion situations.

Generally, we treat uniform circular motion problems as force problems. We follow all the same steps for analyzing force problems:

- draw a free-body diagram
- choose an appropriate coordinate system
- split forces into components
- apply Newton's second law, once for each direction

The only difference is that the acceleration, which is directed toward the center of the circle, has a special form:

$$a_c = \frac{v^2}{r}.$$ We call this the **centripetal acceleration**.

    Let's practice our method on a few different circular motion examples.

1.

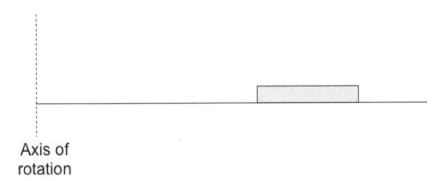

Axis of
rotation

Consider the situation of an object placed on a turntable and rotating with the turntable. Draw a free-body diagram for the object when it is to the right of the center of the turntable.

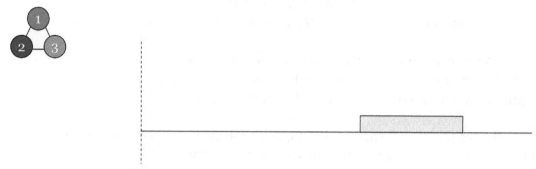

Axis of
rotation

Check with the other members of your group. If necessary, amend your free-body diagram here.

Now apply Newton's second law twice, once for each direction.

Under what condition will the object start to slide?

Water bucket: A water bucket of mass m is being whirled in a vertical circle of radius r at constant speed v. What is the minimum speed required so that the water remains in the bucket without falling out?

Your free-body diagram	Group free-body diagram

Analyze the appropriate situation.

Roller coaster: You're on a roller coaster traveling very fast at the bottom of the loop-the-loop, and you feel pressed into your seat. Effectively you feel like you have an apparent weight that is larger than usual. Why?

Your free-body diagram	Group free-body diagram

Analyze the appropriate situation.

The acceleration of the Earth, associated with its orbit around the Sun.

If you drop an object near the surface of the Earth, it experiences an acceleration with a magnitude of about 9.8 m/s². Take a wild guess at the magnitude of the Earth's acceleration because of its approximately circular orbit around the Sun: ____

Now, we'll calculate it, but we need to know some things, like how fast the Earth is traveling.

The radius of the Earth's orbit around the Sun is:

The time it takes the Earth to travel around the Sun is:

Show how you can use the radius and the time to get the Earth's speed. (Do you feel like you're moving at that speed, by the way??)

Knowing the speed, now we can find the centripetal acceleration. Calculate the centripetal acceleration of the Earth that is associated with its orbit around the Sun.

What force, applied by what, is associated with this acceleration?

Ball on a string: A ball of mass *m*, on a string, is being whirled in midair in a horizontal circle of radius *r* at constant speed *v*. What is the relationship between the angle of the string (measured from the vertical) and *v* and *r*?

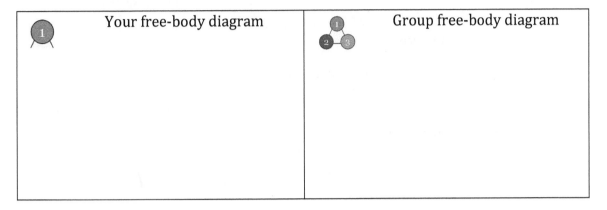

Find the relationship between the angle of the string (measured from the vertical) and *v* and *r*.

Carnival ride: The Gravitron is a carnival ride in which people are stuck to a vertical wall as it rotates (in some versions they remove the floor). Determine the minimum speed the rider can travel so as not to slide down the wall.

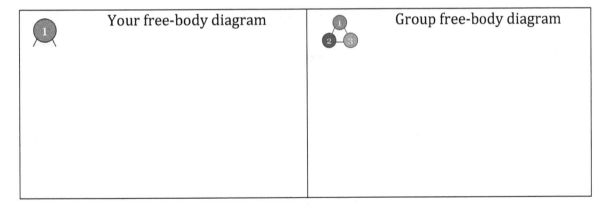

Solve for the minimum speed the rider can travel so as not to slide down.

The equation for momentum:

Momentum is a   [ ] scalar   [ ] vector

The definition of impulse, and its connection to momentum:

When is momentum conserved?

When is momentum not conserved?

Imagine a collision between two carts on a track. For such a collision, what do we expect to be conserved?

[ ] the momentum just of cart 1, as well as the momentum just of cart 2

[ ] the momentum of the two-cart system

[ ] both of the above   [ ] neither of the above

For a collision, momentum conservation is a consequence of one of Newton's laws of motion. Elaborate on this:

Impulse and Momentum
Let's extend our methods by bringing in new ideas, including momentum.

Learning goals – by the end of this section you should be able to
- Translate from a force-vs.-time graph to momentum
- use momentum conservation ideas to analyze physical situations.

One useful idea is that an object's change in momentum is equal to the area under the curve of the net force versus time graph. Let's explore this idea a little.

A cart with a mass of 500 g is subjected to the net force shown in the graph. At t = 0 the cart has a velocity of 4.0 m/s in the positive x-direction.

F_x (N)

Create a table showing the cart's momentum at 1-second intervals.

Time (s)	Momentum (kg m/s)
0	
1	
2	
3	
4	
5	
6	
7	

Plot a graph of the cart's momentum as a function of time.

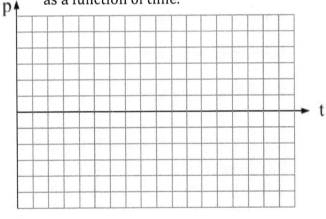

p

t

What is the cart's velocity at t = 5.0 s?

Does the cart change direction at any time? If so, at what time(s)?

Happy ball / sad ball

In the demonstration of the happy ball and the sad ball, predict which ball will knock over the block of wood. Justify your prediction.

Analyze the two ball-block collisions from the point of view of momentum conservation. Let's start with the case of the happy ball.

Before the collision	**After the collision**

Momentum of happy ball = $+mv$ Momentum of happy ball =

Momentum of block =   Momentum of block =

Total momentum =   Total momentum =

Now analyze the collision involving the sad ball.

Before the collision	**After the collision**

Momentum of sad ball = $+mv$ Momentum of sad ball =

Momentum of block =   Momentum of block =

Total momentum =   Total momentum =

Two carts are placed side-by-side, at rest on a horizontal track. A spring in one cart is then released. It pushes against the second cart, causing the carts to move in opposite directions. Is momentum conserved in this situation? Explain your answer.

A cart of mass m and velocity v to the left collides with an identical cart that is initially at rest. After the collision the two carts stick together and move as one unit. What is their velocity after the collision?

Pre-session: Center of mass

Direct link: https://www.youtube.com/embed/4MBAa1cNNUo

What is the equation for the x-coordinate of the center of mass?

Give an example of why the center of mass is a useful concept.

What is the equation for the velocity of the center of mass?

The velocity of the center of mass will be unchanged when ...

In a collision between two parts of a system, the system's center of mass velocity will be

[ ] reversed   [ ] unchanged   [ ] smaller after the collision than before

Center-of-mass

Learning goals – by the end of this section you should be able to
- Calculate the position of a system's center-of-mass
- Use the center-of-mass concept to help analyze a physical situation.

$$\text{Center of mass: } x_{cm} = \frac{m_1 x_1 + m_2 x_2 + \dots}{m_1 + m_2 + \dots}$$

1. Three balls are placed on the meter stick. **The meter stick has negligible mass.** The 1.0 kg ball is at the 0 cm mark; the 3.0 kg ball is at the 80 cm mark, and the 2.0 kg ball is at the 90 cm mark. Calculate the position of the center-of-mass of this system.

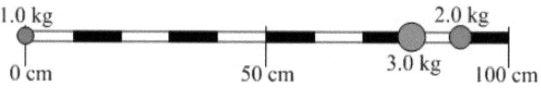

2. A placekicker kicks a football that would otherwise travel a total distance of 50 m. At the top of the flight, the ball explodes into two equal pieces. One piece stops completely, and then falls straight down. Ignore air resistance. The distance from the kicker at which the second piece lands is:

3. Three balls are placed so that there is one ball at each corner of a triangle. You have complete control over the mass of each ball (it can be anything from zero to infinity). As you adjust the various masses, the position of the system center-of-mass changes, but the system center-of-mass is always found

_____ .

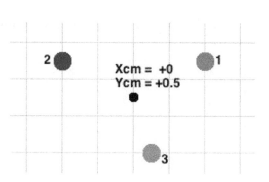

A guy and a canoe

Center of mass: $x_{cm} = \dfrac{m_1 x_1 + m_2 x_2 + \ldots}{m_1 + m_2 + \ldots}$

A man with a mass of 90 kg sits at the end of a 2.4-meter long canoe, which has a mass of 30 kg. **Everything is at rest, initially.** How far from the center of the canoe is the center of mass of the man-canoe system?

The man then gets up and moves to the opposite end of the canoe. How far from the center of the canoe is the center of mass of the man-canoe system now?

How far does the canoe move (relative to a fixed point on the shore)? We'll try to answer this question by drawing pictures.

Sketch the original situation:

Sketch the second situation:

What needs to stay in the same place, when the man changes position? Based on that, how far does the canoe move, relative to a fixed point on the shore?

More Momentum Concepts

A cart with a mass m and a velocity of 2.0 m/s to the right collides with a stationary cart of mass 5m. The collision is completely inelastic (the carts stick together after the collision).

 What is the velocity of the two-cart system after the collision?

 Check your answer with the other members of your group.

A cart with a mass m and a velocity of 2.0 m/s to the right collides with a stationary cart of mass 5m. The collision is elastic (kinetic energy is conserved).

 What is the velocity of each cart after the collision?

 Check your answer with the other members of your group.

Here, we'll do some comparisons between impulse and work.

The impulse equation is:

Impulse is a   [ ] vector   [ ] scalar

The net work equation is:

Work is a   [ ] vector   [ ] scalar

The net work equation has an angle in it. That angle is ...

Impulse is the change in _____

Net work is the change in _____

Impulse is the area under the _____ graph.

Net work is the area under the _____ graph.

Another way to determine the net work is to first find the work done by individual forces. The work done by a force *F* is given by:

How do you then calculate the net work?

The equation for gravitational potential energy is:

Work and Kinetic Energy

Learning goals – by the end of this section you should be able to
- Translate from a force-vs.-position graph to kinetic energy
- Compare and contrast impulse/momentum and work/energy concepts.

We've looked at the connection between force and momentum. Let's look at the analogous connection between force and kinetic energy.

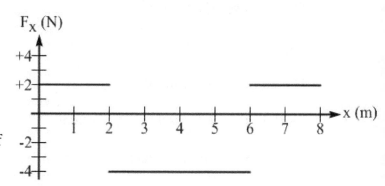

F_x (N)

An object's change in kinetic energy is equal to the area under the curve of the net force versus position graph. Let's explore this idea a little.

A cart with a mass of 500 g is subjected to the net force shown in the graph. When the cart is at x = 0, the cart has a velocity of $4\sqrt{2}$ m/s in the positive x-direction.

Make a table of the cart's kinetic energy as a function of position.

Plot a graph of the cart's kinetic energy as a function of position.

Position (m)	Kinetic Energy (J)
0	
1	
2	
3	
4	
5	
6	
7	
8	

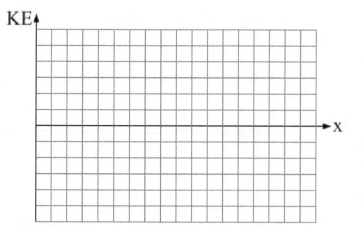

KE

What is the cart's speed when it passes through x = 4.0 m?

What minimum kinetic energy must the cart have at x = 0 to reach x = +8 m?

Comparing impulse and work

Two disks are initially at rest. The mass of disk B is two times larger than that of disk A. The two disks then experience equal net forces F. These net forces are applied for **the same amount of time**. After the net forces are removed, how does the momentum of disk A compare to the momentum of disk B?

Disk A, mass m

Disk B, mass $2m$

[ ] Disk A has more momentum

[ ] Disk B has more momentum

[ ] The disks have equal momenta

What about kinetic energy?

[ ] Disk A has more kinetic energy [ ] Disk B has more kinetic energy

[ ] The disks have equal kinetic energies

The experiment is repeated, but now the forces are applied over **the same distance**. After the net forces are removed, how does the momentum of disk A compare to the momentum of disk B?

[ ] Disk A has more momentum [ ] Disk B has more momentum

[ ] The disks have equal momenta

What about kinetic energy?

[ ] Disk A has more kinetic energy [ ] Disk B has more kinetic energy

[ ] The disks have equal kinetic energies

A car traveling at an initial speed v on a flat road comes to a stop in a distance L once the brakes are applied. Assuming the friction force that brings the car to rest does not change, how far does the car travel once the brakes are applied if the initial speed is $2v$?

Pre-session: Conservation of energy

Direct link: https://www.youtube.com/embed/l9I5veKubew

What is the definition of a **conservative force**? Give an example of such a force.

What is the definition of a **non-conservative force**? Give an example of such a force.

Write out the five-term conservation of energy equation:

Define each of the five terms.

Apply energy conservation to solve the problem of a ball launched from the top of a high cliff – what is the speed of the ball just before impact with the ground?

1. Write out the five-term equation:

2. Cross out any terms that are zero:

3. Make appropriate substitutions for the remaining terms

4. Solve for the speed just before impact

Does the final speed depend on the ball's mass?   [ ] Yes   [ ] No

Does the final speed depend on the launch angle?   [ ] Yes   [ ] No

Name an important piece of information you can't get just from the energy analysis above.

Conservation of Energy
Energy conservation is a powerful idea – let's practice using it.

Learning goals – by the end of this section you should be able to
- Apply a step-by-step method for analyzing a physical situation from the perspective of conservation of energy
- Understand the usefulness and limitations of the energy approach.

Virtually any energy problem can be solved by applying the general energy-conservation equation:

$$U_i + K_i + W_{nc} = U_f + K_f$$

Where U_i is the initial potential energy; K_i is the initial kinetic energy, W_{nc} is the work done by non-conservative forces (such as friction); U_f is the final potential energy; and K_f is the final kinetic energy.

Use the energy-conservation equation to solve the following problems.
Use g = 10 N/kg to keep the calculations simple.

A small block with a mass of 1.0 kg is dropped from rest from a height of 1.8 m above the ground. Neglecting air resistance, find the speed of the block just before it hits the ground. Don't put in any numbers until step 4! Use variables until that point.

Step 1 – Define a zero level for gravitational potential energy. Where is your zero level?

Step 2 – Write down the full five-term energy conservation equation (see above) and then cross out all the terms that are zero. Briefly justify why the terms you crossed out are zero.

Step 3 – Expand the terms that remain, using relations like $U = mgh$ and $K = \dfrac{1}{2}mv^2$.

Step 4 – Solve for what you're trying to solve for. First come up with an expression for the speed in terms of variables and then plug in the numbers given above.

If we doubled the mass of the block, how would that affect the final speed?

If we doubled the height from which we let the block go, how would that affect the final speed?

Let's apply the same step-by-step analysis to solve this problem. A block with a mass of 1.0 kg is released from rest from the top of a ramp that has the shape of a 3-4-5 triangle. The ramp measures 1.8 m high by 2.4 m wide, with the hypotenuse of the ramp measuring 3.0 m. What is the speed of the block when it reaches the bottom, assuming there is no friction between the block and the ramp?

Step 1 - Define a zero level for gravitational potential energy. Where is your zero level?

Step 2 – Write down the full five-term energy conservation equation (see above) and then cross out all the terms that are zero. Briefly justify why the terms you crossed out are zero.

Step 3 – Expand the terms that remain, using relations like $U = mgh$ and $K = \frac{1}{2}mv^2$.

Step 4 – Solve for what you're trying to solve for. First, come up with an expression for the speed in terms of variables and then plug in the numbers given above.

It turns out that we can't neglect friction for the block, because we find that the block's speed at the bottom of the ramp is 2.0 m/s less than the value we calculated above. Use the energy-conservation equation to find a numerical value for the work done by friction on the block.

Use the definition of work, $W = F \Delta r \cos\phi$, (where ϕ is the angle between the force $\vec{F}$ and the displacement $\Delta \vec{r}$) to come up with an expression for the work done by friction in this situation.

Put your results together to solve for the coefficient of kinetic friction associated with the interaction between the block and the ramp. Express your answer as a ratio of integers (e.g., $\mu_K = \frac{3}{7}$).

Virtually any energy problem can be solved by applying the general energy-conservation equation: $U_i + K_i + W_{nc} = U_f + K_f$

A small block of mass m is placed on a track and released from rest from a point 2.2 m above the bottom of the track. The track is frictionless except for a horizontal part that is 2.0 m wide. After passing once through this horizontal part the block reaches a maximum height of 1.7 m above of the bottom of the track on the right.

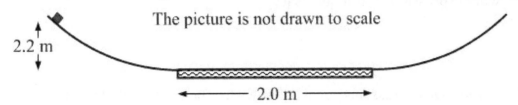

The picture is not drawn to scale

2.2 m

2.0 m

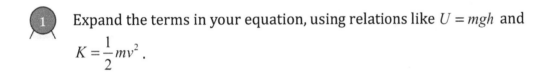

Write down the full five-term energy conservation equation and then cross out all the terms that are zero. Briefly justify why the terms you crossed out are zero.

Check your result with the other members of your group. Re-write the energy expression here, if necessary.

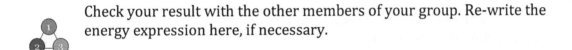

Expand the terms in your equation, using relations like $U = mgh$ and $K = \frac{1}{2}mv^2$.

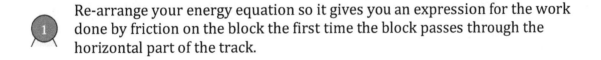

Re-arrange your energy equation so it gives you an expression for the work done by friction on the block the first time the block passes through the horizontal part of the track.

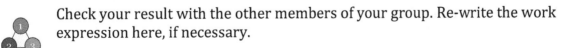

Check your result with the other members of your group. Re-write the work expression here, if necessary.

Use the definition of work (involving force and distance) to determine an expression for the work done by friction on the block the first time the block passes through the horizontal part of the track.

 Check your result with the other members of your group. Re-write the work expression here, if necessary.

 Set your two expressions equal and solve for the coefficient of kinetic friction between the block and the part of the track where there is friction.

 Check your result with the other members of your group. Show your corrected method here, if necessary.

 Discuss the following questions with your group:

How many times will the block completely pass through the horizontal part of the track?

Where does the block finally come to rest?

Pre-session: Power, and energy vs. force

Direct link: https://www.youtube.com/embed/kqK4MdlYoIw

Power is the rate at which _____ is done.

Equation for power:

Unit for power:

The connection between the power unit and the energy unit is:

Equation for power that involves velocity:

What is the approximate average power output of a human being? Show the steps.

We now have two methods we can use to analyze physical situations:
(1) apply energy conservation, or
(2) combine a force analysis with constant acceleration equations.

Which would you use to solve these problems? (You can choose both answers, if appropriate.)

Find the speed of a block sliding down a frictionless incline.

     [ ] Energy   [ ] Force / acceleration

Find the time it takes the block to slide down that incline.

     [ ] Energy   [ ] Force / acceleration

Find the maximum speed of a ball swinging back and forth on a string, pendulum style.

     [ ] Energy   [ ] Force / acceleration

More energy conservation
Virtually any energy problem can be solved by applying the general energy-conservation equation: $U_i + K_i + W_{nc} = U_f + K_f$

Learning goals – by the end of this section you should be
- More comfortable analyzing situations from the perspective of energy.

Loop-the-loop
A block of mass m is placed on a loop-the-loop track and released from rest from a height h above the bottom of the frictionless track. If the loop has a radius R what is the minimum value of h such that the block makes it completely around the loop without losing touch with the track?

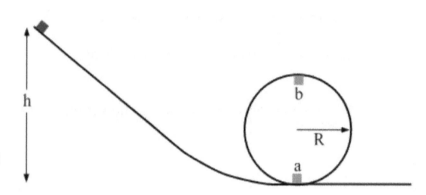

 Step 1 – Define a zero level for gravitational potential energy. Where is your zero level?

 Step 2 – Write down the full five-term energy conservation equation (see above) and then cross out all the terms that are zero. Briefly justify why the terms you crossed out are zero.

 Check your result with the other members of your group. Re-write the energy expression here, if necessary.

 Step 3 – Expand the terms that remain, using relations like $U = mgh$ and $K = \dfrac{1}{2}mv^2$.

Energy is not enough to solve for the minimum height. We also need to do a circular-motion analysis to help us find the minimum kinetic energy the block needs to have at the most critical part of the track. Where is this most critical point? (i.e., where is it in most danger of falling off?)

 Check your answer with the other members of your group.

 Do a circular motion analysis, analyzing the forces on the block at this critical point to find an expression that should help you solve the energy equation above.

 Check your answer with the other members of your group.

 Combine your force result with the energy expression from the previous page, to solve for the minimum value of *h* needed for the block to make it around the loop without falling off.

 Check your answer with the other members of your group. Show your corrected answer here, if necessary.

Pre-session: Combining momentum and energy

Direct link: https://www.youtube.com/embed/W6kV1vBYJWo

Write out the **elasticity** equation:

Classifying collisions – fill in the table

Type of collision	Kinetic energy	Elasticity

In general, which of these is conserved in a collision?

[ ] The total system momentum   [ ] The total system kinetic energy
    [ ] Both   [ ] Neither

A basketball of mass $3m$ and velocity v collides with a baseball of mass m and velocity $-v$. The collision is elastic. Follow the process below to solve for V_f and v_f, the velocities of the basketball and baseball after the collision.

Write out, and then simplify, the equation you get from momentum conservation.

Write out, and then simplify, the equation you get from elasticity.

Put the two equations together to solve for the final velocities.

Combining Energy and Momentum
Sometimes we need to apply both energy conservation and momentum conservation.

Learning goals – by the end of this section you should be able to
- Analyze situations that involve both energy conservation and momentum conservation, and know when to apply which conservation law.

   Work on this together as a group.

A ballistic pendulum is a device used to measure the speed of a bullet. A bullet of mass m is fired at a block of wood (mass M) hanging from a string. The bullet embeds itself in the block, and causes the combined block plus bullet system to swing up a height h. What is v_0, the speed of the bullet before it hits the block?

First, is it correct to set the bullet's initial kinetic energy equal to the final gravitational potential energy of the block plus bullet? If not explain why not, and describe the method you would use instead to find the answer.

Find an expression for the initial speed of the bullet.

Colliding pendula

Two balls hang from strings of the same length. Ball A, with a mass of 4 kg, is swung back to a point 0.8 m above its equilibrium position. Ball A is released from rest and swings down and hits ball B. After the collision ball A rebounds to a height of 0.2 m above its equilibrium position, and ball B swings up to a height of 0.05 m. Use g = 10 N/kg.

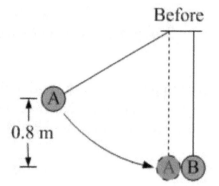

Before

 What is the speed of ball A just before the collision?

 Check your answer, and your method, with the other members of your group.

 What is the speed of ball A just after the collision?

 What is the speed of ball B just after the collision?

 Calculate the mass of ball B.

Check your answer, and your method, for the mass of ball B with the other members of your group. If necessary, show your corrected work here.

What type of collision is this?

[ ] super-elastic [ ] elastic   [ ] inelastic [ ] completely inelastic

Justify your answer (two different ways, if possible).

Energy in a hopper popper

Figure out what the launch speed of the popper is, and how much energy is stored in the popper when you turn it inside out.

Estimate the average power output of a human being, based on a daily energy consumption of 2500 "calories" – those are actually kilocalories, and we can approximate that there are 4000 J / kcal. Round your calculations off so you can estimate the number without a calculator.

A disk is spinning. A particular point on the disk has a particular velocity at a particular instant in time. At that same instant, how many other points on the disk have that same velocity?

[ ] None   [ ] 1   [ ] 2   [ ] 3   [ ] All of them

At that same instant, how many other points on the disk have the same **angular velocity** as the original point?

[ ] None   [ ] 1   [ ] 2   [ ] 3   [ ] All of them

The angular acceleration of the disk is connected to...

[ ] the centripetal acceleration of a point on the disk

[ ] the tangential acceleration of a point on the disk

What is the connection between...

... the angle θ a disk rotates through and the distance x travelled by a point on the disk?

... the disk's angular speed ω and the tangential velocity of a point on the disk?

... the disk's angular acceleration α and the tangential acceleration of a point on the disk?

Rotation
Today marks the start of the several days we'll spend looking at rotating systems. Much of what we learning previously can be applied to such systems.

Learning goals – by the end of this section you should be able to
- Recognize and make use of the parallels between straight-line motion and rotational motion.
- Apply the rotational kinematics equations to solve problems when the angular acceleration is constant.

Consider a system that consists of a large block tied to a string wrapped around the outside of a rather large pulley. The pulley has a radius of 2.0 m. When the system is released from rest, the block falls with a constant acceleration of 0.5 m/s², directed down.

What is the speed of the block after 4.0 s?

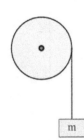

How far does the block travel in 4.0 s?

Plot a graph of the speed of the block as a function of time, up until 4.0 s.

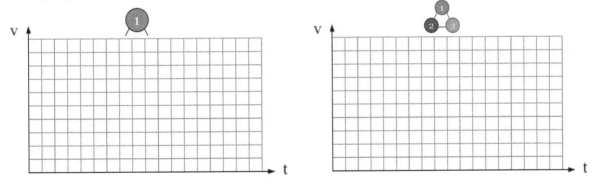

Also, plot the speed of a point on the pulley that is on the outer edge of the pulley, 2.0 m from the center. Finally, plot the speed of a point on the pulley that is 1.0 m from the center.

Rotational Kinematics
Motion with constant (angular) acceleration

We will often analyze the rotational motion of a rotating system using constant-acceleration equations that have the same form as the one-dimensional constant-acceleration equations we used before.

Straight-line motion equation	Analogous rotational motion equation
$v = v_i + at$	$\omega = \omega_i + \alpha t$
$x = x_i + v_i t + \dfrac{1}{2}at^2$	$\theta = \theta_i + \omega_i t + \dfrac{1}{2}\alpha t^2$
$v^2 = v_i^2 + 2a(\Delta x)$	$\omega^2 = \omega_i^2 + 2\alpha(\Delta\theta)$

Spinning a bike wheel: While fixing the chain on your bike, you have the bike upside down. Your friend gives the front wheel, which has a radius of 30 cm, a spin. You observe that the wheel has an initial angular velocity of 2.0 rad/s, and that the wheel comes to rest after 50 s.

Assume that the wheel has a constant angular acceleration. Either by using the equations above, or by using other means, determine how many revolutions the wheel makes.

Check your answer against that of the other members of your group.

A ferris wheel

Straight-line motion equation	Analogous rotational motion equation
$v = v_i + at$	$\omega = \omega_i + \alpha t$
$x = x_i + v_i t + \dfrac{1}{2}at^2$	$\theta = \theta_i + \omega_i t + \dfrac{1}{2}\alpha t^2$
$v^2 = v_i^2 + 2a(\Delta x)$	$\omega^2 = \omega_i^2 + 2\alpha(\Delta\theta)$

You are on a ferris wheel that is rotating at the rate of 1 revolution every 2π seconds. The operator of the ferris wheel decides to bring it to a stop, and puts on the brake. The brake produces a constant acceleration of -0.10 radians/s².

Use this space to organize what you know, and maybe draw a picture.

 (a)   If your seat on the ferris wheel is 4.0 m from the center of the wheel, what is your speed when the wheel is turning at a constant rate, before the brake is applied?

 As a group, see if you can come up with another way to find your speed.

(b) How long does it take before the ferris wheel comes to a stop?

 Compare your answer, and your method, with the other members of your group. Make note of other ways to solve the problem, if there are any.

 (c) How many revolutions does the wheel make while it is slowing down?

 Compare your answer, and your method, with the other members of your group. Make note of other ways to solve the problem, if there are any.

 (d) How far do you travel while the wheel is slowing down?

 Compare your answer, and your method, with the other members of your group. Make note of other ways to solve the problem, if there are any.

Pre-session: Torque

Direct link: https://www.youtube.com/embed/a8ofWd0j51Y

Torque is the rotational equivalent of _____.

A force is a push or a pull. A torque is a _____.

A net torque acting on an object at rest will cause the object to _____ .

The equation for the magnitude of a torque is:

The direction of the torque associated with the force F,
shown here, about an axis through the left end of the rod is

    [ ] clockwise   [ ] counterclockwise

Describe a method you can use to find the torque direction.

Three ways to find torque – here, find the torque
applied by F_T, the tension in the string, acting on
the rod, about the axis of rotation.

Method 1 (describe, show the result, and draw)

Method 2 (describe, show the result, and draw)

Method 3 (describe, show the result, and draw)

Torque
Torque is essentially the rotational equivalent of force. Let's spend some time figuring out what exactly that statement means.

Learning goals – by the end of this section you should be able to
- Apply torque to compare different rotational situations.
- Understand the three equivalent ways to calculate torque.

If you have ever opened a door you have a basic understanding of torque. How quickly the door opens depends on the force you apply to the door, where on the door you exert the force, and in what direction you exert the force. All of these factor into the torque you exert on the door.

Consider the following situations involving a revolving door.

Situation 1 – overhead view of a revolving door

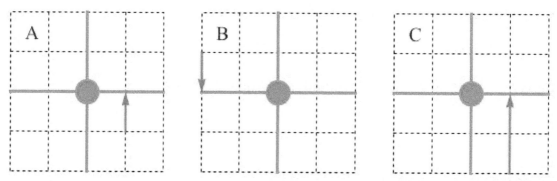

A - a force *F* is applied perpendicular to the door halfway between the center and an edge.

B - a force *F* is applied perpendicular to the door at one edge.

C - a force 2*F* is applied perpendicular to the door halfway between the center and an edge.

 Rank these based on the magnitude of the torque experienced by the door, from largest to smallest. This is equivalent to ranking the situations based on their angular accelerations (i.e., ranking them by how effective the force is at getting the door to spin).

Compare your answers to those of the other members of your group.

Situation 2 – overhead view of a revolving door

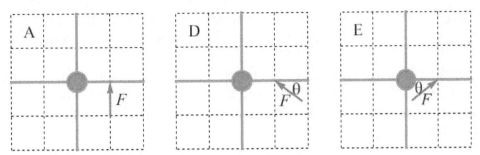

In each case the magnitude of the force is the same and the force is applied to the door halfway between the center and one edge.

A - the force is directed perpendicular to the door.

D - the angle between the force and the door is 30°, with a component toward the center.

E - the angle between the force and the door is also 30°, with a component away from the center.

 Rank these situations based on the magnitude of the torque experienced by the door, from largest to smallest.

 Compare your answers to those of the other members of your group.

Situation 3 – overhead view of a revolving door

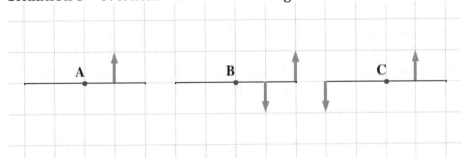

 Rank these situations based on the magnitude of the torque experienced by the door, from largest to smallest.

 Compare your answers to those of the other members of your group.

The magnitude of the torque resulting from applying a force F at a distance r from an axis of rotation is: $\tau = r\,F\sin\theta$. The angle θ represents the angle between the line of the force and the line the distance r is measured along. Torque is a vector, so its direction is also important. We usually handle the direction separately, however.

Consider three equivalent ways of finding the torque applied by a particular force. A rod of length L is hinged at one end and held horizontal by means of a string attached to the other end. Find the torque associated with the tension in the string, about an axis passing through the hinge.

Method 1 – use the whole force and the angle. First, in what direction is the torque from the tension? Clockwise or counterclockwise?

Identify the three factors in the torque equation above.
What is F in this case?

What is r in this case?

What is the angle between F and r?

Apply the torque equation to find torque.

Method 2 – Split the force into vertical and horizontal components. Sketch this.

Vertical component
What is F in this case?

What is r in this case?

What is the angle between F and r?

Horizontal component
What is F in this case?

What is r in this case?

What is the angle between F and r?

Apply the torque equation to find torque. Add the two torques together as vectors (account for direction with + and – signs) to find the net torque.

Method 3 – The lever-arm method. Draw a line from the hinge to the string so the line meets the string at a 90° angle. Measure r along this line. Sketch this.
What is F in this case?

What is r in this case?

What is the angle between F and r?

Apply the torque equation to find torque.

Pre-session: Static equilibrium

Direct link: https://www.youtube.com/embed/MuesopkJNQA

If an object is in static equilibrium it means the object is at rest and stays at rest.

What are the two conditions that must be satisfied for an object to remain in static equilibrium? (Express in words and in equations.)

A uniform beam with a weight of 12 N rests on two supports. On the picture of the beam on the right, sketch the free-body diagram of the beam.
Write out the force equation for this situation. Can you solve it?

Let's say that support B is a distance d from the midpoint of the rod, and A is $2d$ from the midpoint.

Take torques around the midpoint – write out the torque equation. What does this tell you about how the support forces compare?

Take torques around the point where support B is – write out the torque equation. Solve for the support force from support A.

Take torques around the point where support A is – write out the torque equation. Solve for the support force from support B.

Static Equilibrium
How do we analyze a system that remains at rest?

Learning goals – by the end of this section you should be able to
- Recognize a static equilibrium situation.
- Write down the conditions that apply in static equilibrium.
- Write one or more torque equations to analyze a static equilibrium situation.
- Combine force equations and torque equations to solve an equilibrium problem.

A long rod that is free to rotate about one end is being held horizontal by an upward force exerted at the opposite end of the rod. This upward force is measured by a spring scale. Sketch the free-body diagram for this situation. What rule(s) does your free-body diagram have to satisfy?

The reading on the spring scale is _____ N. This shows us the force exerted at the far end of the rod. What is the weight of the rod itself?

The place where the spring scale is attached will now be moved toward the hinge. What, if anything, happens to the reading on the spring scale? Explain why.

Draw a new free-body diagram for the situation in which the spring scale is quite close to the hinge, ¼ of the way along the rod as measured from the hinge. Predict what the spring scale will read.

Static Equilibrium

At equilibrium $\vec{a} = 0$ *and* $\vec{\alpha} = 0$*. This means that* $\sum \vec{F} = 0$ *and* $\sum \vec{\tau} = 0$*.*

A rod with a length L and a mass m is attached to a wall by means of a hinge at the left end. The rod's mass is uniformly distributed along its length. A string will hold the rod in a horizontal position; the string can be tied to one of three points, lettered A-C, on the rod. The other end of the string can be tied to one of three hooks, numbered 1-3, above the rod. This system could be a simple model of a broken arm you want to immobilize with a sling. The rod represents the lower arm, the hinge represents the elbow, and the string acts as the sling.

For each case below, draw a line (and only one line) from one lettered point to one numbered hook representing the string you would use to achieve the desired objective. If you think it is impossible to achieve the objective explain why.

How would you attach a string so the rod is held in a horizontal position but the hinge exerts no force at all on the rod?

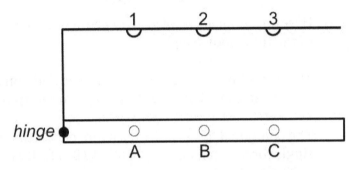

How would you attach a string so the rod is held in a horizontal position while the force exerted on the rod by the hinge has no horizontal component, but has a non-zero vertical component directed straight up?

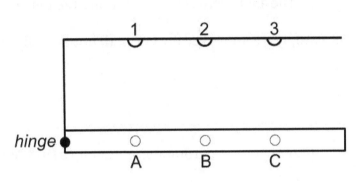

How would you attach a string so the rod is held in a horizontal position while the force exerted on the rod by the hinge has no vertical component, but has a non-zero horizontal component?

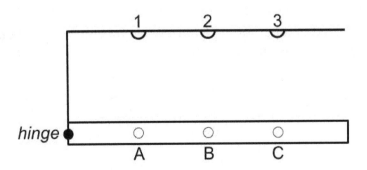

Static equilibrium applied to us

Static equilibrium can be applied to the human body in all sorts of ways, such as how you can keep your arm at rest while holding a heavy object in your hand. Another good example is the human spine, which, in this example, we will model as a rigid rod.

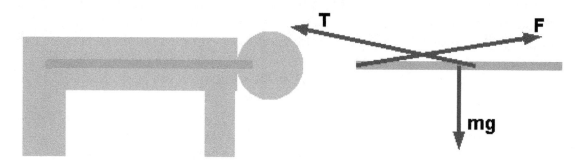

Here, *mg* is the force of gravity acting on the upper body, which we take to be 65% of the total body weight.

There is a force *F* applied by the lower body on the upper body at the tailbone. This force is at a small angle with respect to the spine.

The combined force of the back muscles acting on the spine we will model as a single force of tension (*T*) applied 10% farther from the tailbone than the *mg* force is. The angle between the spine and the tension force is 12 degrees.

Apply Newton's second law both horizontally and vertically.

Choose an axis of rotation, and set up a sum-of-torques equation.

For a person with a total body weight of 600 N, mg is about 400 N. Solve for the values of *F* and *T*.

Rotational dynamics

Learning goals – by the end of this section you should be able to
- Apply Newton's second law for rotation.
- Recognize the parallel between Newton's second law and the second law for rotation.

Now we'll consider a non-equilibrium situation involving a system we looked at before – a rod, hinged at one end, held in a horizontal position by a string. The string holding the rod is cut, and the rod starts to swing down toward the ground.

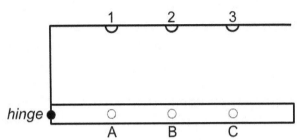

The rotational inertia of a uniform rod rotating about an axis through one end is $I = \frac{1}{3}mL^2$. Another useful equation in this situation is that $\alpha = \frac{a_T}{r}$, where a_T is the tangential acceleration. Remember that the magnitude of a torque is $\tau = r\,F\sin\theta$. Immediately after the string is cut, when the rod is still horizontal, find:

(a) The acceleration of the rod's center-of-mass.
(b) The magnitude and direction of the force applied to the rod by the hinge.

Step 1 – Sketch a free-body diagram of the rod.

Step 2 – Apply Newton's Second Law, $\Sigma \vec{F} = m\vec{a}$.

Step 3 – Apply Newton's Second Law for Rotation, $\Sigma \vec{\tau} = I\vec{\alpha}$.

Step 4 – Solve the problem.

Table of rotational inertias

Note that one thing you should just know is that the rotational inertia of an object in which all its mass is the same distance (or, essentially the same distance) r from the point of rotation is given by:

$$I = mr^2$$

Examples would include a ball being whirled on the end of a string, the Earth as it rotates around the Sun, and a person on a large merry-go-round.

For other objects, which have their masses distributed at different distances from the axis of rotation, we generally provide the value from the table below.

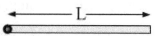

Rod rotating about an axis through one end, perpendicular to the rod.

$$I = \frac{1}{3}ML^2$$

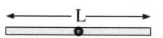

Rod rotating about an axis through the middle, perpendicular to the rod.

$$I = \frac{1}{12}ML^2$$

Solid disk or cylinder about an axis through the middle, perpendicular to the plane of the disk.

$$I = \frac{1}{2}MR^2$$

Solid sphere about an axis through the center.

$$I = \frac{2}{5}MR^2$$

Thin ring about an axis through the middle, perpendicular to the plane of the ring.

$$I = MR^2$$

Hollow sphere about an axis through the center.

$$I = \frac{2}{3}MR^2$$

These results are calculated by splitting the object into tiny pieces, and adding up the mr^2 from each piece to get the total rotational inertia (this is done using calculus).

Atwood's machine, re-visited
Let's do a better job analyzing Atwood's machine

An Atwood's machine is a device that has two objects connected by a string that passes over a pulley. Assume $M > m$, and that the pulley is a solid disk with a mass m_p.

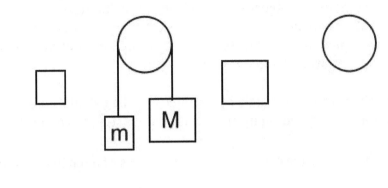

Sketch free-body diagrams showing the forces acting on the objects of mass M and m and on the pulley.

For mass M, show which direction you're taking to be positive. Apply Newton's second law to obtain a relationship between M, a (the acceleration), and the forces acting on mass M.

For mass m, show which direction you're taking to be positive. Apply Newton's second law to obtain a relationship between m, a (the acceleration), and the forces acting on mass m.

For the pulley, show which direction you're taking to be positive. Apply Newton's second law to obtain a relationship between m_p, α (the angular acceleration), and the torques acting on the pulley.

See if you can combine your three equations to find an expression for the acceleration a in terms of g, m, M, and m_p.

Rotational Inertia – Calculating rotational inertia

Learning goals – by the end of this section you should be able to
- Calculate the rotational inertia of a system, about a particular axis.

Inertia is our measure of how difficult it is to change an object's motion. For straight-line motion, an object's inertia is given by its mass. For rotational motion the rotational inertia depends on mass, how the mass is distributed, and even on what axis the object is rotating about. The symbol for rotational inertia is I, and it represents the rotational analog of mass.

In general, the rotational inertia of an object can be calculated by splitting the object into small pieces and adding up the rotational inertia of each piece using the equation:

$$I = \sum_i m_i r_i^2$$, where m_i represents the mass of a particular piece and r_i

represents the distance of that piece from a particular axis of rotation. In practice we rarely use this method, and instead look up expressions for moments of inertia in a table, but it's nice to get some idea of how the method is used so we know where the expressions come from.

Let's try this for a system consisting of three blocks, each with a mass of $\dfrac{m}{3}$. The

blocks are placed on a light rod of length L with a block at each end and one block in the center. Find an expression for the system's rotational inertia about an axis through the center of the rod.

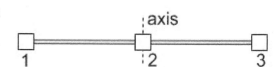

We can find the rotational inertia of an individual block using $I = mr^2$.
Find the rotational inertia of:

Block 1:

Block 2:

Block 3:

Compare your answers to those of the other members of your group.

Add the three inertias together to find an expression for the rotational inertia of the system for this particular axis of rotation.

Let's repeat the process, but this time the axis of rotation is through one end of the rod.

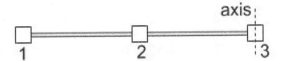

 Find the rotational inertia of:

Block 1:

Block 2:

Block 3:

 Compare your answers to those of the other members of your group.

Add them together to find an expression for the rotational inertia of the system for this particular axis of rotation.

You can also find the answer using the **parallel-axis theorem**.

$$I = I_{CM} + mh^2$$

The rotational inertia about an axis is the rotational inertia of the parallel axis through the center-of-mass, plus the mass of the system multiplied by the square of the distance between the new axis and the axis through the center-of-mass.

Starting from the result on the previous page (that's the rotational inertia about the axis through center-of-mass), add on the appropriate amount based on the total mass of the system and how far the axis was shifted from the center-of-mass. Does the total agree with what you calculated above?

Pre-session: Angular momentum

Direct link: https://www.youtube.com/embed/hOFz04X7-LM

First, let's review linear momentum (we just referred to this as momentum, earlier).

Equation for linear momentum: p =

Linear momentum is a   [ ] vector   [ ] scalar

If no net force acts on a system, the system's linear momentum is _____

If there is a net force, the linear momentum changes. Δp =

You should see the parallels between linear momentum and angular momentum.

Equation for angular momentum: L =

Angular momentum is a   [ ] vector   [ ] scalar

If no net **torque** acts on a system, the system's angular momentum is _____

If there is a net **torque**, the angular momentum changes. ΔL =

Use angular momentum conservation to explain why a figure skater rotates faster when she is spinning and then moves her arms closer to her body.

Another parallel:

The connection between torque and force: τ =

The connection between angular momentum and linear momentum: L =

Direction of angular momentum – if you watch a movie on a screen that shows an object spinning clockwise, its angular momentum is actually directed

[ ] out of the screen   [ ] into the screen

A Rotational Collision
Analyzing a rotational collision is a lot like analyzing a collision in one-dimension.

Learning goals – by the end of this section you should be able to
- Apply conservation of angular momentum to analyze a rotational collision.

Review: For a one-dimensional collision, we use the equation $p = mv$ and, as long as there is no net external force acting, we can say that the total system momentum before the collision is equal to the total system momentum after the collision.

Here's the equivalent for a rotational collision: we use the equation $L = I\omega$ and, as long as there is no net external torque acting, we can say that the total system angular momentum before the collision is equal to the total system angular momentum after the collision.

Example: A uniform solid disk of mass M and radius R is rotating about its center at a constant angular speed of 6.0 rad/s.

A piece of wood of mass $3M$ and length $2R$ is carefully dropped from rest onto the disk. The two objects then rotate together with the center of the rod at the center of the disk.

What is their angular speed after this collision?

What type of collision is this?

[ ] super-elastic   [ ] elastic   [ ] inelastic   [ ] completely inelastic

Write down the expression for the angular momentum before the collision.

Write down the expression for the total angular momentum of the system after the collision.

Set these two equations equal to each other (angular momentum is conserved) and solve for the angular speed after the collision.

Pre-session: Rotational Kinetic Energy

Direct link: https://www.youtube.com/embed/_GFVjYjStm0

Kinetic energy for an object that is just moving in a straight line: KE =

Kinetic energy for an object that is just rotating: KE =

Kinetic energy for an object that is both moving in a straight line and rotating:

KE =

Now, let's summarize some of the parallels between linear motion and rotational motion.

Variable	Linear variable	Rotational variable	Connection
Position			
Velocity			
Acceleration			
Produces Acceleration			
Inertia			
Momentum			

Rotational Kinetic Energy
Analyzing a system that includes rotational kinetic energy.

Learning goals – by the end of this section you should be able to
- Incorporate rotational kinetic energy into an energy-conservation analysis.

The equation for rotational kinetic energy is analogous to the $\frac{1}{2}mv^2$ equation we're used to for translational kinetic energy. Rotational kinetic energy is $K_{rot} = \frac{1}{2}I\omega^2$. This can be directly incorporated into our usual five-term energy equation.

An Atwood machine consists of a large block of mass $7m$ tied to a smaller block of mass $3m$. The string passes over a frictionless pulley that is a uniform solid disk of mass $4m$. After being released from rest, how fast is the large block moving after dropping 1.0 m?

We could answer this using forces and torques (which we did already) but let's try energy. Start with the usual conservation of energy equation: $K_i + U_i + W_{nc} = K_f + U_f$.

Cross out all the terms that are zero in this equation.

Write out expressions for the remaining terms. Remember to account for both translational kinetic energy and rotational kinetic energy, if appropriate. Keep everything in terms of variables, at first. Note that the pulley is a uniform solid disk, with a rotational inertia of $I = \frac{1}{2}MR^2$.

Find an expression for the speed of each block after they have moved 1.0 m, starting from rest. Note that, because the string does not slip, we can use $\omega = v/R$. Solve for the speed numerically.

Pre-session: Simple harmonic motion

Direct link: https://www.youtube.com/embed/9jeyU4Rac48

What is the equation for the force exerted by an ideal spring? (This equation is known as Hooke's law.)

Name, define, and give the units for, the constant in the equation.

What is the implication of the minus sign in the equation?

What is the equation for the elastic potential energy of a spring?

Use energy conservation to determine the maximum speed of a block of mass m on a spring of spring constant k, released from rest a distance A from equilibrium. Show work.

Springs and Simple Harmonic Motion
Let's think about springs, and the motion of an object that is connected to a spring.

Learning goals – by the end of this section you should be able to
- Determine the magnitude and direction of the force applied by a spring.
- Incorporate elastic potential energy into an energy conservation analysis.
- Define angular frequency and know how to use it.

We have spent a lot of time analyzing forces, so let's talk about springs from that perspective. We have dealt with cases in which forces are constant. A spring, however, exerts a varying force that depends on how much it is stretched or compressed.

We will use a simple model in which we assume the spring force is opposite in direction to, and proportional to, the stretch or compression. This model is known as:

Hooke's Law: $\vec{F}_{spring} = -k\vec{x}$,   where k is known as the spring constant.

A spring hangs vertically down from a support. When you hang a 100-gram mass from the bottom end of the spring and stop any motion of the system, the spring is stretched by 10 cm. Determine the spring constant. Hint: sketch a free-body diagram.

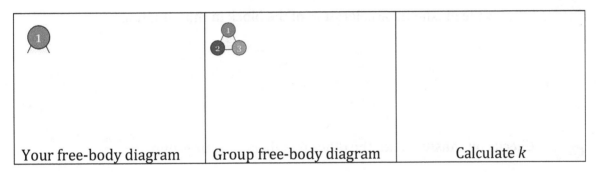

| Your free-body diagram | Group free-body diagram | Calculate k |

If we hang another 100-gram mass on the spring the spring stretches further. The additional stretch is:

[ ] another 10 cm   [ ] less than 10 cm   [ ] more than 10 cm

Check your answer with the other members of your group.

Springs store energy, so we can define what we call an elastic potential energy, which we can build directly into our conservation of energy analysis:

Elastic potential energy: $U_e = \dfrac{1}{2}kx^2$.

A block with a mass of 1.0 kg is released from rest on a frictionless incline. At the bottom of the incline, which is 1.8 m vertically below where the block started, the block slides across a horizontal frictionless surface before encountering a spring that has a spring constant of 100 N/m. What is the maximum compression of the spring? Hint: use energy conservation.

 Check your answer with the other members of your group.

 What is the maximum acceleration of the block in this situation?

 Check your answer with the other members of your group.

A system experiencing simple harmonic frequency has a natural oscillation frequency. For an object of mass m oscillating on a spring of spring constant k the angular frequency is given by:

$$\omega = \sqrt{\frac{k}{m}}.$$

The angular frequency is related to T, the period of oscillation, and f, the frequency, by:

$$\omega = \frac{2\pi}{T} = 2\pi f .$$

In all situations below, we start the oscillations by displacing the object a certain distance from the equilibrium position and releasing it from rest.

We have an object on a spring that oscillates at a particular frequency. We then change either the mass or the spring constant and let the object oscillate again. By comparing graphs of the object's position, velocity, and acceleration vs. time after the change to those before the change, can you tell what was changed?

 If you can tell, which one was changed in the situation shown in class?

[ ] the mass   [ ] the spring constant   [ ] we can't tell

Justify your answer:

Check your answer and your explanation with the other members of your group.

What if we plot the energy (potential, kinetic, and total) of the system before and after the change instead? By comparing graphs of the object's energy vs. time after the change to those before the change, can you tell what was changed?

If you can tell, which one was changed in the situation shown in class?

[ ] the mass   [ ] the spring constant   [ ] we can't tell

Justify your answer:

Check your answer and your explanation with the other members of your group.

Pre-session: Simple harmonic motion II

Direct link: https://www.youtube.com/embed/oU7dzQAhbdc

A block of mass m is attached to an ideal spring of spring constant k. $x = 0$ corresponds to the equilibrium position. If the block is released from rest from $x = +A$ at $t = 0$ s, what is the equation giving the block's position as a function of time?

What is the corresponding equation for the block's velocity as a function of time?

What is the corresponding equation for the block's acceleration as a function of time?

What is the equation for the angular frequency, ω, for a block on a spring?

The simple pendulum (a ball on a string)

Sketch a free-body diagram for a ball on a string, first hanging motionless, and then as it is passing through the equilibrium position while oscillating.

           At rest   Passing through equilibrium

Sketch a free-body diagram for a ball on a string, after displacing the ball to the left. Use a coordinate system parallel and perpendicular to the string.

What's the equation for the angular frequency, ω, for a simple pendulum?

The role of time in simple harmonic motion.
Let's spend some time thinking about time in simple harmonic motion.

Learning goals – by the end of this section you should be able to
- Be comfortable using the time equations in simple harmonic motion.
- Know how to analyze a simple pendulum.

Energy tells us nothing about time. However, if Hooke's Law can be applied (i.e., if an object feels a force directed opposite to, and proportional to, its displacement from equilibrium), and if mechanical energy is conserved, we say that the resulting motion is simple harmonic motion. Such motion can be described by an equation of the form: $\bar{x}(t) = A\cos(\omega t)$, where A is the amplitude (maximum displacement from equilibrium) of the motion and ω is known as the angular frequency. For an object of mass m oscillating on a spring of spring constant k we have:

$$\omega = \sqrt{\frac{k}{m}}.$$ This is a measure of the rate at which the object oscillates.

The angular frequency is related to the period of oscillation by: $\omega = \dfrac{2\pi}{T}$.

The velocity and acceleration of the object are given by:

$$\bar{v} = -A\omega\sin(\omega t) = -v_{max}\sin(\omega t) \quad \text{and} \quad \bar{a} = -A\omega^2\cos(\omega t) = -a_{max}\cos(\omega t) = -\omega^2\bar{x}$$

Lesson 1: Don't use the time equations unless you really need them. An object of mass m is attached to a spring that has a spring constant k and released from rest a distance A from equilibrium. The object oscillates on a frictionless horizontal surface.

(1) Find the maximum speed reached by the object during the resulting oscillations. Hint: because this problem asks you to relate position and speed, try using energy conservation.

 Check your answer with the other members of your group.

Lesson 2: Know how to use the time equations when you do need them. You generally have to work in radians. An object attached to a spring is pulled a distance A from the equilibrium position and released from rest. It then experiences simple harmonic motion with a period T. The time taken to travel between the equilibrium position and a point A from equilibrium is T/4.

 How much time is taken to travel between points A/2 from equilibrium and A from equilibrium? Assume the points are on the same side of the equilibrium position, and that mechanical energy is conserved. Let's say the object is A from equilibrium at t = 0, so the equation $\bar{x}(t) = A\cos(\omega t)$ applies. Now just solve for the time t when the object is A/2 from equilibrium.

Check your answer with the other members of your group.

The Simple Pendulum
The pendulum is another classic example of harmonic motion.

An example of a simple pendulum is a ball on the end of a string. The ball can be treated as if all of its mass is concentrated at one point, and the mass of the string can be neglected.

Draw a simple pendulum so the string is at an angle θ with respect to the vertical.

You release the simple pendulum (the ball on the string) from the position given above. Sketch a free-body diagram of the ball showing all the forces acting on the ball just after you release it. Is there a net force acting on the ball?

		Is there a net force acting on the ball? [ ] Yes [ ] No
Your free-body diagram	Group free-body diagram	

Sketch a second free-body diagram that applies when the ball is passing through the equilibrium position (when the string is vertical). Is there a net force acting on the ball?

		Is there a net force acting on the ball? [ ] Yes [ ] No
Your free-body diagram	Group free-body diagram	

What do you think the pendulum's oscillation frequency depends on?

We'll experiment a little with some pendula in class – what does the frequency depend on?

Pre-session: Fluids (The Buoyant Force)

Direct link: https://www.youtube.com/embed/zPb3FPq_qGw

Define a **fluid**:

We'll have a more specific definition of the buoyant force later, but we'll start with a general definition. The **buoyant force** is ...

A wooden block with a weight of 100 N floats (in equilibrium) exactly 50% submerged in a particular fluid. The upward buoyant force exerted on the block by the fluid ...

[ ] has a magnitude of 100 N   [ ] has a magnitude of 50 N

[ ] depends on the density of the block   [ ] depends on the density of the fluid

[ ] depends on both these densities

Draw a free-body diagram of the block to support your answer above.

In the video, we draw an analogy between the buoyant force and the _____ force.

The Buoyant Force
Let's work the buoyant force into our standard treatment of forces.

Learning goals – by the end of this section you should be able to
- Incorporate the buoyant force into a free-body diagram and a force equation.
- Use Archimedes' principle to solve problems.

How do we define a fluid in physics? Provide some examples.

A block of weight $mg = 45.0$ N has part of its volume submerged in a beaker of water. The block is partially supported by a string of fixed length. When 80.0% of the block's volume is submerged, the tension in the string is 5.00 N.

What is the magnitude of the buoyant force acting on the block? Hint: try drawing a free-body diagram of the block and applying Newton's Second Law.

Your free-body diagram	Group free-body diagram	Find the buoyant force.

Water is steadily removed from the beaker, causing the block to become less submerged. The string breaks when its tension exceeds 35.0 N. What percent of the block's volume is submerged at the moment the string breaks?

Your free-body diagram	Group free-body diagram	Find the buoyant force, then how much submerged.

After the string breaks and the block comes to a new equilibrium position in the beaker, what percent of the block's volume is submerged?

Your free-body diagram	Group free-body diagram	Find the buoyant force, then how much submerged.

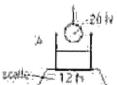

In Figure A, a 20 N ball is supported by a string. It hangs over a beaker of fluid that sits on a scale. The scale reading is 12 N. In Figure B the ball is completely submerged in the fluid. In Figure C the ball is exactly half submerged. In Figure D the string has been cut and the ball rests on the bottom of the beaker.

In figure A, what is the tension in the string?

[ ] 0   [ ] 5 N   [ ] 10 N   [ ] 15 N   [ ] 20 N   [ ] 32 N

In figure B, what is the buoyant force on the ball?

[ ] 0   [ ] 5 N   [ ] 10 N   [ ] 15 N   [ ] 20 N   [ ] 32 N

In figure B, what is the scale reading?

[ ] 7 N   [ ] 12 N   [ ] 17 N   [ ] 22 N   [ ] 27 N   [ ] 32 N

In figure C, what is the buoyant force on the ball?

[ ] 0   [ ] 5 N   [ ] 10 N   [ ] 15 N   [ ] 20 N   [ ] 32 N

In figure C, what is the tension in the string?

[ ] 0   [ ] 5 N   [ ] 10 N   [ ] 15 N   [ ] 20 N   [ ] 32 N

In figure C, what is the scale reading?

[ ] 7 N   [ ] 12 N   [ ] 17 N   [ ] 22 N   [ ] 27 N   [ ] 32 N

In figure D, what is the buoyant force on the ball?

[ ] 0   [ ] 5 N   [ ] 10 N   [ ] 15 N   [ ] 20 N   [ ] 32 N

In figure D, what is the scale reading?

[ ] 7 N   [ ] 12 N   [ ] 17 N   [ ] 22 N   [ ] 27 N   [ ] 32 N

Pre-session: Pressure

Direct link: https://www.youtube.com/embed/QvBb4i00hOU

Write out **Archimedes' principle** in words.

Write out **Archimedes' principle** as an equation:

Write the equation for mass density:

Define **specific gravity**.

Write the equation for **pressure**:

What is the unit we use for pressure? Show how this unit is related to units we have used before.

Write out the equation relating the pressure of two points in a static fluid. Define the terms.

Explain the connection between pressure and the buoyant force.

Pressure
Understanding fluids requires an understanding of pressure.

Learning goals – by the end of this section you should be able to
- Explain from a microscopic perspective how a fluid exerts pressure on an object.
- Appreciate the tremendous forces that can arise from atmospheric pressure.

Have a look at the picture of the three points in the J-shaped tube. The tube is closed at the top on the right, and open to the atmosphere on the left.

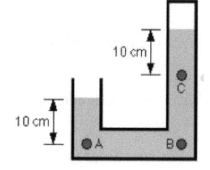

  Rank these points based on the pressure at the points, from largest to smallest. Justify your ranking.

 Check your ranking with the other members of your group.

We'll do a couple of demonstrations involving atmospheric pressure. One involves crushing a can with atmospheric pressure. To help understand this, draw two diagrams. Each diagram should show the forces being exerted on the outside of the can and the inside of the can.

Diagram 1: Can open to the atmosphere.	Diagram 2: Air removed from inside the can.

Estimate the force that the atmosphere exerts on one side of the can.

Another good atmospheric pressure demonstration is the vacuum cannon, in which a ping pong ball is propelled down an evacuated tube by atmospheric pressure. Estimate the speed at which the ball comes out of the tube assuming the following:

Atmospheric pressure is 1×10^5 Pa.

Pressure inside the tube is 0.

Mass of the ball is 4 g.

Area of the ball is 1.2×10^{-3} m^2.

Zero friction or resistance for the ball.

The tube length is 2 m.

Find the force exerted on the ball by the air.

Either find the ball's acceleration and use a constant-acceleration equation, or use the work-energy relationship, to find the ball's speed at the end of the tube.

Direct link: https://www.youtube.com/embed/R56yAwJ2VDI

What are the four conditions an ideal fluid must satisfy for us to be able to analyze it using fairly simple equations?

1.

2.

3.

4.

Write out the **continuity equation**:

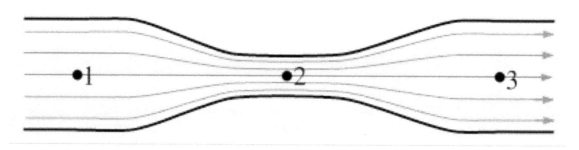

An ideal pipe is flowing through a horizontal pipe that has a constriction in the middle (in other words, it's narrower in the middle than elsewhere). At which point is the fluid pressure the greatest?

[ ] point 2   [ ] points 1 and 3   [ ] equal at all these points

Write out **Bernoulli's equation**:

Fluid Dynamics
How do we deal with flowing fluids? In some cases, flowing fluids are too complicated to analyze. In others we apply a simple model with two equations:

Learning goals – by the end of this section you should be able to
- Understand and make use of the two equations we apply to flowing fluids.
- Explain the concept of applying energy conservation to a flowing fluid.

Continuity equation (based on mass conservation): $A_1 v_1 = A_2 v_2$.

In a tube the continuity equation relates the flow speed and pipe area at one point to their values at another point.

Bernoulli's equation (energy conservation): $\rho g y_1 + \dfrac{1}{2}\rho v_1^2 + P_1 = \rho g y_2 + \dfrac{1}{2}\rho v_2^2 + P_2$

What are the units of the six terms in Bernoulli's equation?

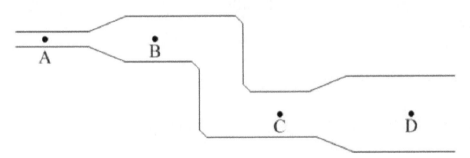

Four points are marked in a pipe containing a fluid. The pipe's cross-sectional area at B is the same as that at C.

 If the fluid is at rest in the tube, rank the four points based on their pressure, from largest to smallest.

 Compare your rankings to those of the other members of your group. Adjust your rankings, if necessary.

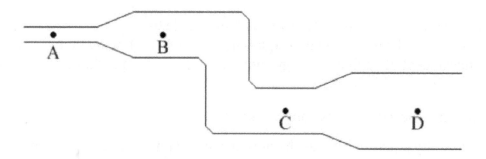

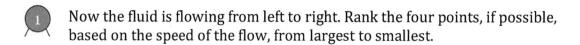

1. Now the fluid is flowing from left to right. Rank the four points, if possible, based on the speed of the flow, from largest to smallest.

Compare your rankings to those of the other members of your group. Adjust your rankings, if necessary.

1. In the situation of the fluid flowing to the right rank the four points based on their pressure, from largest to smallest.

Compare your rankings to those of the other members of your group. Adjust your rankings, if necessary.

Would your rankings change if the fluid was flowing right to left instead?

There's a hole in my bucket...

A cylinder full of water stands upright on a table. The cylinder has three holes in its sides that are initially covered with tape. One hole is ¼ of the way down from the top, another is ½ way down, and the third is ¾ of the way down. When the tape is removed, water shoots horizontally out of the holes and lands on the table. Assume the cylinder is maintained so that it is always full to the top with water.

Step 1 – Sketch a diagram of the situation.

Predict which hole shoots the water farthest horizontally on the table.

[ ] The hole closest to the top   [ ] The hole halfway down

[ ] The hole closest to the bottom   [ ] It's a three-way tie

Briefly justify your answer:

Check your prediction with the other members of your group. If you change your prediction after discussion, explain why.

Let's analyze the situation generally, using a cylinder with height H that is filled with water, and a hole that is a distance h from the top. Once the water emerges from the tube we have a projectile motion situation, which we know how to handle. Let's figure out the speed with which the water emerges from the cylinder.

Step 2 – To find the speed with which the water emerges from the hole, write down Bernoulli's equation. Let's take point 2 to be the point outside the cylinder where the fluid emerges from the hole. Pick a convenient point inside the cylinder (any point will actually work) to be point 1.

Step 3 – Define a zero for gravitational potential energy. Cross out all terms that are zero in the equation, and see if any terms cancel out.

Copy over your equation from step 3 on the previous page.

Step 4 – Bring in the continuity equation to do something about v_1. Feel free to make a reasonable assumption about how the area of the hole compares to the cross-sectional area of the cylinder.

Step 5 – Solve for v_2, the speed at which the water emerges from the hole.

Step 6 – Use projectile motion methods to solve for the horizontal distance traveled by the water. Challenge: see if you can use your equation to predict which hole will shoot the water farthest horizontally.

When we account for viscosity, we're accounting for the fact that most real fluids

flow with _____ to the flow.

The viscosity of a fluid can be measured by shearing it (not like a sheep) – moving a plate of area A over the top of the fluid, making the top layer move. Other layers also move, but not as much as the top layer. In this case, the drag force when the plate moves at a constant velocity v is given by (also, define all the variables):

What are the units for viscosity?

What is the value for the viscosity of room temperature water?

About how much more viscous is honey compared to water?

Viscosity

Learning goals – by the end of this section you should be able to
- Apply the drag force equation to understand the motion of an object falling through a fluid at terminal velocity.
- Explain how an ultracentrifuge works.

Drag force on a ball moving through a viscous fluid: $F_d = 6\pi\eta r v$

The ball has a radius r and a speed v, and the fluid has viscosity η.

If you release a ball from rest in a fluid, it will accelerate until it reaches a terminal (constant) velocity. Sketch the free-body diagram for the ball when it is traveling at its terminal velocity. Hint: there are three forces acting on it. Apply Newton's second law to get an equation relating the forces.

Starting from your force equation, see if you can verify one, or both, of these expressions for the terminal velocity.

$$v_t = \left(1 - \frac{\rho_{fluid}}{\rho_{ball}}\right)\frac{mg}{6\pi\eta r} \qquad\qquad v_t = \left(\rho_{ball} - \rho_{fluid}\right)\frac{2gr^2}{9\eta}$$

If you double the radius and keep the density the same, how will the terminal velocity change?

Physics of a centrifuge

Why doesn't blood separate into its constituents? For example, why don't red blood cells sink to the bottom? We will calculate how long it would take for a red blood cell to move 1 cm at its terminal velocity. For simplicity, approximate the red blood cell as a sphere.

Average density of blood = 1060 kg/m^3

Density of red blood cell = 1125 kg/m^3;   Radius of red blood cell ≈ 3.5 x 10^{-6} m

Mass of red blood cell ≈ 27 x 10^{-15} kg;   Viscosity of blood ≈ 3.5 x 10^{-3} Pa s

What is the blood cell's terminal velocity?

$$v_t = \left(1 - \frac{\rho_{fluid}}{\rho_{ball}}\right) \frac{mg}{6\pi\eta r}$$

At this speed, how long does it take the red blood cell to travel 1 cm?

To isolate red blood cells, blood samples are placed in a centrifuge. Spinning a sample in a circle at high speed is like placing it in a stronger gravitational field. Let's say that instead of g, the blood in the centrifuge acts like it is in a gravitational field of 5000 g. How does this affect the time it takes a red blood cell to travel 1 cm through the sample?

Circular motion review: Let's say that to achieve an acceleration of 5000 g, the centrifuge operates at 5000 rpm. At what radius should the blood sample be placed?

Pre-session: Life at Low Reynolds number

Direct link (to a paper, not a video):
https://www.gwu.edu/~phy21bio/Reading/Purcell_life_at_low_reynolds_number.pdf

Page 1: Water is unusual in that most liquids have viscosities _____ that

of water.   [ ] higher than   [ ] lower than   [ ] about the same as

Pages 2-3: At very low Reynolds number (such as that experienced by tiny

creatures in water), motion is dominated by   [ ] inertia   [ ] viscosity

.

Page 4: Which two of these are productive methods of locomotion at low Reynolds

number?   [ ] stiff oars   [ ] flexible oars   [ ] scallop shell   [ ] corkscrew

Page 6: E. coli swim by doing what with their flagella?

[ ] waving them up and down   [ ] rotating them

Page 8: The propulsive efficiency of tiny creatures moving around in a low Reynolds

number environment is approximately   [ ] 100%   [ ] 10%   [ ] 1%

Page 11: What is meant by the phrase "If you don't swim that far, you haven't gone

anywhere?"

Propelling an *E. coli*
Note: this exercise is based on an activity from the University of Maryland NEXUS project

Learning goals – by the end of this section you should be able to
- Apply ideas from fluids to understand the motion of an *E. coli*.

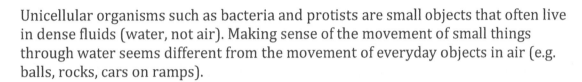

Unicellular organisms such as bacteria and protists are small objects that often live in dense fluids (water, not air). Making sense of the movement of small things through water seems different from the movement of everyday objects in air (e.g. balls, rocks, cars on ramps).

Consider the following conversation between two biology students:

> Liz: I work in a microbiology lab, and in our experiments we are often manipulating the motion of objects suspended in a fluid. In such a fluid, to get objects to move at a constant speed, we have to apply a force that is proportional to the *velocity* of the object.

> Jack: That can't be right. In physics, I learned that force is equal to the mass of the object times the *acceleration* of the object.

In your group, first make sure you understand each perspective. What is Liz saying about the relationship between speed and force? What is Jack saying?

In the remainder of this problem we'll use the example of an *E. coli* swimming through water to further understand whether the perspectives of these two students are consistent or contradictory.

E. coli move by rotating their flagella to move through the fluid. The fluid pushes back on them, by Newton's third law. We will call this force of the fluid on the flagella of the *E. coli* "the applied force", F_{app} (it is present because the *E. coli* moves its flagella). This is the force that moves the *E. coli* forward. The *E. coli* also feels a resistive force, in the opposite direction. This is a viscous force (associated with the viscosity of the fluid), with a magnitude given by

$$F_{viscous} = \beta R v$$

The viscous force is proportional to the speed and the effective radius of the object and opposite to the velocity. (The "effective radius" of an object depends somewhat on its shape and its size, but for an object that is not too stretched out, we can take it to be some average radius of the object.)

Draw a free-body diagram that shows the forces acting on the *E. coli* as it moves. Assume the *E. coli* is moving horizontally through the water.

Compare your diagram to those of the other students in your group. If you change your diagram, draw the new version here.

If the *E. coli* is moving at a constant velocity v_T, what is the applied force? Write an equation that could give a value for F_{app}. How will this value depend on the shape and size of the *E. coli* (*i.e.*, its mass (m) and effective radius (R))?

Solve the equation that you found for v_T. How does v_T depend on the applied force? How would v_T change if the *E. coli* pushed harder on the fluid?

It's not very realistic for the *E. coli* to start at constant velocity. Suppose the *E. coli*, starting from rest, starts to move and comes fairly quickly to its constant velocity, v_T. You can assume that the applied force is constant, meaning that the *E. coli* is pushing on the fluid with a constant force. Sketch graphs of x, v, a, F_{net}, F_{app}, and $F_{viscous}$, and explain what is happening to each variable and why. (You might find it helpful to consider what would happen if the *E. coli* applied a different force, and draw more than one curve on your graph.)

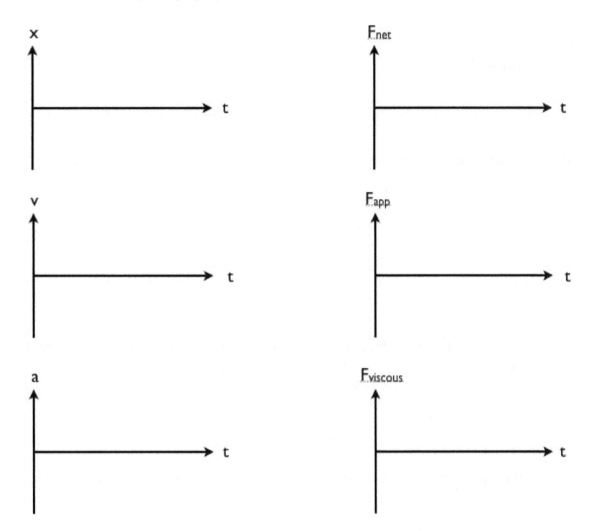

Now, return to the original question. Are the perspectives of Liz and Jack contradictory? Consistent? If they are consistent, explain how they can be reconciled. If they are contradictory, explain the inconsistencies. (Regardless of the position you take, please address both the arguments of Liz and Jack.)

Define **temperature**.

Define **internal energy**.

What are the equations we use to convert from Fahrenheit to Celsius, and vice versa?

What do you have to be careful of when using an equation involving temperature, T, that you don't have to worry about when the equation involves ΔT?

What's the equation we use for linear thermal expansion? Define the variables.

What's the equation we use for volume thermal expansion?

Thermal Expansion, and Specific Heat
Objects change size when their temperature changes; and heat produces either a change in temperature or a change in phase

Linear thermal expansion: $\Delta L = L_i\, \alpha\, \Delta T$ or, equivalently, $L = L_i\, (1 + \alpha\, \Delta T)$

α is the linear thermal expansion coefficient, which depends on the material;

L is the new length, L_i is the original length, and ΔL is the change in length;

ΔT is the change in temperature.

Strips of two different metals, which have the same length at room temperature, are bonded together to form a bimetallic strip. When the bimetallic strip is heated it bends into a circular arc. Why?

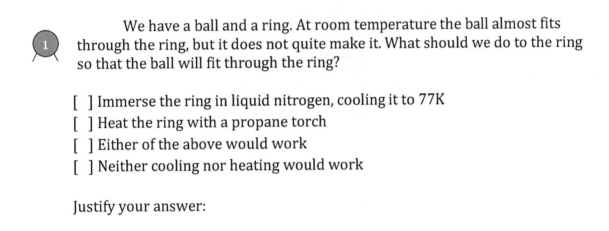

We have a ball and a ring. At room temperature the ball almost fits through the ring, but it does not quite make it. What should we do to the ring so that the ball will fit through the ring?

[ ] Immerse the ring in liquid nitrogen, cooling it to 77K
[ ] Heat the ring with a propane torch
[ ] Either of the above would work
[ ] Neither cooling nor heating would work

Justify your answer:

Check your answer with the other members of your group. If you change your answer after discussion, explain why.

What actually happens? Why?

The amount of heat, Q, required to change the temperature of an object by an amount ΔT is given by:

$$Q = mc\,\Delta T,$$

m is the mass of the object, and c is the **specific heat**, which depends on the material.

The amount of heat, Q, associated with a phase change is:

$$Q = mL,$$

m is the mass of the object, and L is the **latent heat**, which depends on the material, and on which phase change it is. L_f is the **latent heat of fusion**, which applies to the solid-liquid phase change, and L_v is the **latent heat of vaporization**, which applies to the liquid-gas phase change.

 100 grams of ice, with a temperature of –10°C, is added to a styrofoam cup of water. The water is initially at +10°C, and has an unknown mass m. If the final temperature of the mixture is 0°C, what is the unknown mass m? Assume that no heat is exchanged with the cup or with the surroundings. Use these approximate values to determine your answer:

The specific heat of liquid water is about 4000 J/(kg °C)
The specific heat of ice is about 2000 J/(kg °C)
The latent heat of fusion of water is about 3 x 10⁵ J/kg

Hint: there is more than one possible answer for m – find the range of possible answers for m.

Heat Transfer

Learning goals – by the end of this section you should be
- Familiar with the three different mechanisms for heat transfer, and be able to use the conduction and radiation equations.

Conduction: A home freezer, with a wall area of 6.0 m², is insulated with a 5.0 cm thickness of styrofoam (k = 0.033 W/(m K)). At what rate must heat be removed to keep the inside of the freezer at -20°C, if the room is at +20°C?

$$\frac{Q}{\Delta t} = k \frac{A}{L} (T_{\mathrm{H}} - T_{\mathrm{C}})$$

Consider a two-layer problem where one layer has twice the thickness and six times the thermal conductivity as the other layer, but the layers have the same area.

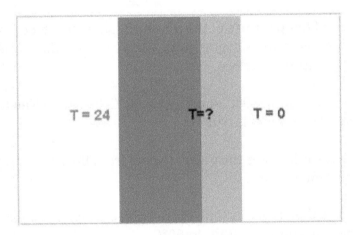

Show how you find the temperature at the interface between the layers, if the temperature on the left face of the thicker layer is 24°C, and the temperature on the right face of the thin layer is 0°C.

Radiation: The Sun emits energy at the rate of 4 x 10²⁶ W, and the Sun's radius is 7 x 10⁸ m. Calculate the surface temperature of the Sun.

$$\frac{Q}{\Delta t} = \varepsilon \sigma A T^4$$

Newton's law of cooling states that the rate at which an object cools is proportional to the temperature difference between the object and the surrounding environment. This leads to an exponential relationship between the object's temperature, as a function of time, as it cools:

$$T_{object} = T_{environment} + (T_{initial} - T_{environment})e^{-\lambda t}$$

$T_{initial}$ is the initial temperature of the object, T_{object} is the object's temperature at a time t after the cooling process begins, and λ is a constant that is related to the decay rate of the temperature.

Let's say that $T_{environment}$ = 20°C, and $T_{initial}$ = 80°C. Based on Newton's law of cooling, which of these statements is true?

[ ] It takes twice as long for the object to cool from 80°C to 40°C as is does for the object to cool from 80°C to 60°C

[ ] It takes more than twice as long for the object to cool from 80°C to 40°C as is does for the object to cool from 80°C to 60°C

[ ] It takes less than twice as long for the object to cool from 80°C to 40°C as is does for the object to cool from 80°C to 60°C

Using Newton's law of cooling, determine the ratio of the two times in the preceding question.

Diffusion and Thermodynamics

Use the Phet simulation called Membrane Channels

Google it, or go here: http://phet.colorado.edu/en/simulation/membrane-channels

Add 50 green particles to the top half (this does not need to be exact, so don't worry if you are off by 1 or 2).

1. Describe the motion of the green particles. Is it random or pre-determined? What could make the particles behave this way?

Add 3 evenly spaced green gated channels to the membrane. Click the "Open Channels" button.

2. The process you are observing is diffusion. Describe this process using the key words **particle, movement, concentration, high, and low**.

Click on the "Clear Particles" button and add 40 green particles to the top half, and add 10 green particles and 30 blue particles to the bottom half. Add 3 blue gates and open those, too (you should already have three open green gates).

3. Make a prediction: after the simulation has run for a while, what will the equilibrium situation be? Specifically, how many green and blue particles do you expect to find on each side of the membrane?

   Top half: # of green particles _____   # of blue particles _____

   Bottom half: # of green particles _____   # of blue particles _____

4. Check your prediction by speeding up the animation to full speed. All gates should be open. When you think equilibrium has been reached, hit the "Pause" button and count the number of blue and green particles on each side. Record the number below:

   Top half: # of green particles _____   # of blue particles _____

   Bottom half: # of green particles _____   # of blue particles _____

   Is this consistent with your prediction? Explain.

Now, close the blue gates but keep the green gates open. Click on the "Clear Particles" button and add 40 green particles to the top half, and add 10 green particles and 30 blue particles to the bottom half.

5. Make a prediction: after the simulation has run for a while, what will the equilibrium situation be? Specifically, how many green and blue particles do you expect to find on each side of the membrane?

Top half: # of green particles _____   # of blue particles _____

Bottom half: # of green particles _____   # of blue particles _____

6. Check your prediction by speeding up the animation to full speed. Only the green gates should be open. When you think equilibrium has been reached, hit the "Pause" button and count the number of blue and green particles on each side. Record the number below:

Top half: # of green particles _____   # of blue particles _____

Bottom half: # of green particles _____   # of blue particles _____

Is this consistent with your prediction? Explain.

We will view this video on diffusion together.
http://www.youtube.com/watch?v=H7QsDs8ZRMI
7. Why do particles diffuse faster in a vacuum?

8. Why do nitrogen oxide particles diffuse faster than bromine particles?

9. How does temperature affect the rate of diffusion? Why do you think this is so? (It has to do with the speed of the particles.)

10. Are there any biological systems you are aware of that utilize diffusion? What are they? What is your understanding of how they work?

Pre-session: The Ideal Gas Law

Direct link: https://www.youtube.com/embed/IC-FnzjuISM

Write out Avogadro's number:

Write out the ideal gas law in terms of R, and state the value of R.

Write out the ideal gas law in terms of k, and state the value of k.

Give two reasons why increasing the temperature in a fixed volume of ideal gas causes the pressure to increase.

1.

2.

When we apply the ideas of kinetic theory (basic physics) to an ideal gas, we come up with a simple equation that tells us what temperature is. What is this equation, and what does it tell us about temperature?

Ideal gases
Kinetic theory involves applying basic ideas of physics (such as impulse, pressure, and kinetic energy) to a container of ideal gas.

Learning goals – by the end of this section you should be able to
- Apply the ideal gas law to physical situations.
- Make use of a P-V diagram to help understand a thermodynamic situation.

Write out the ideal gas law:

In units of J / (mol K) what is the value of R, the universal gas constant?

Express the ideal gas law in terms of N, the number of gas molecules, instead of n, the number of moles of gas.

One of the major results of kinetic theory is that we get some insight into the connection between the average kinetic energy (K_{av}) of the gas molecules in a container of ideal gas and the absolute temperature of that ideal gas:

$$K_{av} = \frac{3}{2}kT \text{ , where } k = 1.38 \times 10^{-23} \text{ J/K is the Boltzmann constant.}$$

This tells us something of fundamental importance. The absolute temperature, which is a macroscopic quantity we measure for the substance as a whole, is a direct measure of something that is going on at the microscopic level, the average kinetic energy of the molecules.

Example Problem: A box of ideal gas consists of light particles and heavy particles (the heavy ones have 16 times the mass of the light ones). Initially all the particles have the same speed. When equilibrium is reached, what will be true?

[ ] All the particles will still have the same speed

[ ] The average speed of the heavy particles equals the average speed of the light particles

[ ] The average speed of the heavy particles is larger than that of the light particles

[ ] The average speed of the heavy particles is smaller than that of the light particles

Explain your answer:

Three identical cylinders are sealed with identical pistons that are free to slide up and down the cylinder without friction. Each cylinder contains ideal gas, and the gas occupies the same volume in each case, but the temperatures differ. In each cylinder the piston is above the gas, and the top of each piston is exposed the atmosphere. In cylinders 1, 2, and 3 the temperatures are 0°C, 50°C, and 100°C, respectively.

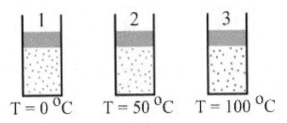

Rank the cylinders in terms of the pressure of the gas, from greatest to least.

Rank the cylinders in terms of the number of moles of gas inside the cylinder.

Sketch a free-body diagram for a piston in one of the cylinders above. Assume the piston has a mass m and that the circular area of the top or bottom of the piston is A. Apply Newton's second law to find the pressure in the cylinder.

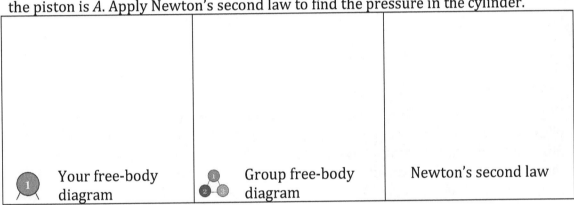

① Your free-body diagram	①②③ Group free-body diagram	Newton's second law

Using m = 5 kg, g = 10 N/kg, A = 0.010 m² and P_{atm} = 1.000 x 10⁵ Pa, find the pressure in the cylinder.

P-V (pressure versus volume) diagrams can be very useful. Consider the P-V graph shown at right.

What are the units resulting from multiplying pressure in kPa by volume in liters?

Rank the four states shown on the diagram based on their absolute temperature, from greatest to least.

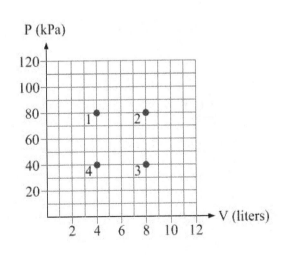

Pre-session: The First Law of Thermodynamics

Direct link: https://www.youtube.com/embed/dpVcdSQYR5I

On this P-V diagram, what are the two curved lines shown here called, and what do they represent?

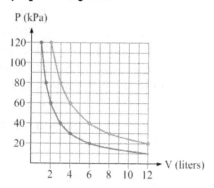

Define thermodynamics.

Write out the first law of thermodynamics, and define all the terms.

The variable W, for work, appears in the first law. You have to be especially careful with the sign of W, because physicists and chemists often use W to represent different (but related) things. Physicists generally say that W is the work done by the system, while chemists often say that W is the work done on the system (therefore flipping the sign of W).

Using the physics definition for W, W is positive when the volume occupied by the gas ...

[ ] increases   [ ] decreases

How do you find work from a P-V diagram?

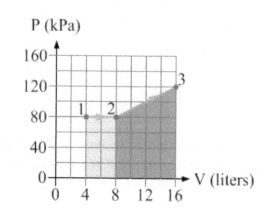

Calculate the work here for...

The 1 → 2 process:

The 2 → 3 process:

Both processes combined:

The First Law of Thermodynamics

Thermodynamics is the study of systems involving energy in the form of internal energy, heat, and work. The First Law relates these three energies.

Learning goals – by the end of this section you should be able to
- Apply energy conservation to thermodynamics (the first law).
- Make use of a P-V diagram to calculate work.

Let's use the following symbols:
Q = heat; W = work; E_{int} = internal energy; ΔE_{int} = change in internal energy.

W represents the work done by the gas. In general W is positive when the volume increases, and negative when the volume decreases.

A sealed can of ideal gas is connected to a cylinder that has a piston that can move up or down. The can is initially at room temperature, but we then place the can in a beaker of hot water. We observe that the piston in the cylinder moves up, doing some work.

Applying the ideas of energy conservation, write out an equation that relates the different types of energy in this process. If you do it right, you should be writing down the First Law of Thermodynamics.

How do we find the work? The work done by the gas in a particular process is the area under the curve on the P-V diagram that corresponds to that process.

How do we find ΔE_{int}, the change in internal energy? The internal energy is directly proportional to temperature, so the change in internal energy is directly proportional to the change in temperature.

The simplest way to write the change in internal energy is $\Delta E_{int} = nC_V \Delta T$.
In this equation, C_V is the heat capacity at constant volume. For a monatomic ideal gas, for instance, $C_V = \frac{3}{2}R$.

We can use $\Delta E_{int} = nC_V \Delta T$ for all processes - the change in internal energy depends only on the change in temperature, and is independent of the process.

That is in contrast to the work, which, as the area under the curve, depends on the exact process involved in moving from one state to another.

To summarize, a good way to express the First Law of Thermodynamics is:
Q = (Area under the curve on the P-V diagram) + $nC_V \Delta T$

For this P-V diagram...

Calculate the work done in the 1 to 2 process.

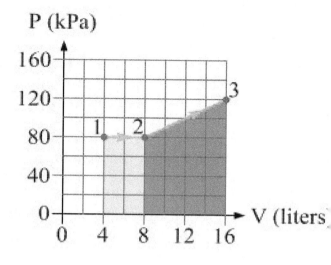

Calculate the work done in the 2 to 3 process.

An ideal gas system, initially in state 1, progresses to some final state by one of three different processes (a, b, or c). Each of the possible final states has the same temperature.

For which process is the change in internal energy larger? Why?

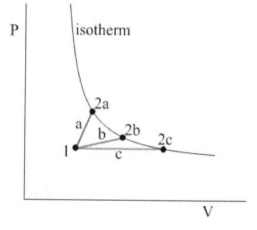

Check your answer with the other members of your group. If you change your answer after discussion, explain why.

For which process is more heat transferred into the ideal gas? Explain.

Check your answer with the other members of your group. If you change your answer after discussion, explain why.

Pre-session: A sample thermodynamics problem

Direct link: https://www.youtube.com/embed/fMUaguA3ZaM

What's the equation we generally apply to gases, for calculating the heat associated with changing temperature? Define all the variables.

For a monatomic ideal gas, what is the value of the heat capacity for constant volume, and the value of the heat capacity for constant pressure?

When we calculate the change in internal energy for an ideal gas, the equation $\Delta E_{int} = nC_V \Delta T$ can be applied ...

[ ] only when the volume is constant   [ ] only for a monatomic ideal gas

[ ] for any process and for any type of ideal gas

For the example problem, show what you know, and show how you can calculate the temperature of the system in state 1.

When we add heat to the gas, at constant pressure, in the example problem (it's a monatomic ideal gas), what percentage of the heat goes to changing the internal energy and what percentage goes to the work done by the gas?

Show how you find the new temperature, after the heat is added at constant pressure.

Thermodynamic Processes

Analyzing thermodynamic processes involves the first law of thermodynamics, the ideal gas law, and the P-V diagram.

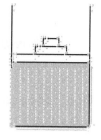

You have some monatomic ideal gas in a cylinder. The cylinder is sealed at the top by a piston that can move up or down, or can even be fixed in place to keep the volume constant. Blocks can be added to, or removed from, the top of the cylinder to adjust the pressure, as necessary.

Starting with the same initial conditions each time, you do three experiments. Each experiment involves the same amount of heat, Q.
A – Add heat Q to the system at constant pressure.
B – Add heat Q to the system at constant temperature.
C – Add heat Q to the system at constant volume.

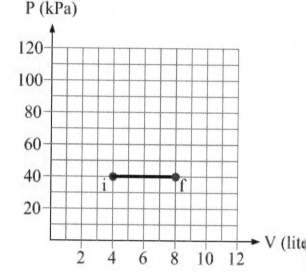

P (kPa)

Sketch these processes on the P-V diagram. The circle with the i beside it represents the initial state of the system. One of the processes is drawn already. Identify which one, and draw the other two.

Rank the experiments, from largest to smallest, based on their final temperatures.

Check your answer with the other members of your group. If you change your answer after discussion, explain why.

Rank the experiments, from largest to smallest, based on the work done by the gas.

Check your answer with the other members of your group. If you change your answer after discussion, explain why.

Each experiment involves the same amount of heat, Q.
A – Add heat Q to the system at constant pressure.
B – Add heat Q to the system at constant temperature.
C – Add heat Q to the system at constant volume.

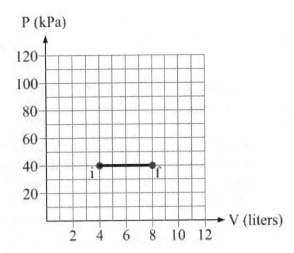

P (kPa)

V (liters)

 Rank the experiments, from largest to smallest, based on their final pressures.

 Check your answer with the other members of your group. If you change your answer after discussion, explain why.

 Find the numerical value of Q.

Find the final pressure in process C.

Find an expression for the final volume in process B.

Cyclic Processes

Learning goals – by the end of this section you should be able to
- Apply the first law and the ideal gas law to thermodynamic cycles.

A thermodynamic system undergoes a three-step process. An adiabatic expansion takes it from state 1 to state 2; heat is added at constant pressure to move the system to state 3; and an isothermal compression returns the system to state 1. The system consists of a diatomic ideal gas with $C_V = 5R/2$. The number of moles is chosen so $nR = 100$ J/K. The following information is known about states 2 and 3. Pressure: $P_2 = P_3 = 100$ kPa Volume: $V_3 = 0.5$ m^3

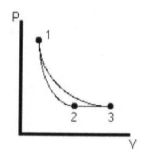

 Calculate the temperature of the system in state 3.

 Check your answer with the other members of your group. If you change your answer after discussion, explain why.

 The system does 20000 J of work in the constant pressure process that takes it from state 2 to state 3. What is the volume and temperature of the system in state 2?

$V_2 =$ $T_2 =$

 Check your answers with the other members of your group. If you change your answer after discussion, explain why.

A thermodynamic system undergoes a three-step process. An adiabatic expansion takes it from state 1 to state 2; heat is added at constant pressure to move the system to state 3; and an isothermal compression returns the system to state 1. The system consists of a diatomic ideal gas with $C_V = 5R/2$. The number of moles is chosen so $nR = 100$ J/K. The following information is known about states 2 and 3.
Pressure: $P_2 = P_3 = 100$ kPa
Volume: $V_3 = 0.5$ m^3

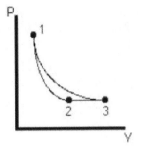

 Complete the table below, filling in the missing numerical values, with units of joules. The total work done by the system in the cycle is –19400 J. Note that there is a way to do this without integrating or using logarithms.

Process	Q	ΔE_{int}	W
1 to 2			
2 to 3			20000 J
3 to 1			
Complete cycle			-19400 J

Step 1: First fill in all the terms that are zero.

Remember that each row in the table satisfies the First Law of Thermodynamics.

Also remember that $\Delta E_{int} = nC_V \Delta T$.

Step 2: Do one calculation, without using logarithms.

Step 3: Now, you should be able to do a Sudoku-like process to fill in the rest of the table.

The Second Law of Thermodynamics
Many useful devices, such as car engines and refrigerators, rely on thermodynamic cycles. These systems also have to obey the second law of thermodynamics

Learning goals – by the end of this section you should be able to
- Explain the flow of energy in a typical heat engine.
- Apply equations in ideal and non-ideal cases to calculate heats and work.

The second law of thermodynamics is all about entropy – in general, for a closed system entropy either stays the same or increases, but does not decrease.

In general, an engine works like this.
Some heat, Q_H, is obtained (such as by burning fuel) at a relatively high temperature.
Some fraction of this goes into doing useful work, W, and the remaining energy must essentially be thrown away, in the form of heat (Q_L) at a lower temperature.

We can express this as an energy conservation equation: $Q_H = W + Q_L$.

The efficiency of the engine is defined as: $e = \dfrac{W}{Q_H} = \dfrac{Q_H - Q_L}{Q_H} = 1 - \dfrac{Q_L}{Q_H}$.

You might think that an efficiency as close to 1 as possible is achievable by building a better engine, but Sadi Carnot (a French engineer) showed that a consequence of the Second Law of Thermodynamics is that even in the best possible case (what we call an ideal engine) the efficiency is:

$e_{ideal} = 1 - \dfrac{T_L}{T_H}$, where T_L is the temperature of the reservoir into which the waste

heat is dumped, and T_H is the temperature at which the heat was obtained by burning gasoline (or equivalent).

(1) For a typical car engine the waste heat is deposited into the atmosphere, which is at a temperature of about 300 K, while the temperature inside the car cylinders might be 1000 K. Calculate the ideal efficiency of such an engine.

 Compare your calculation with the other members of your group.

Heat Engines

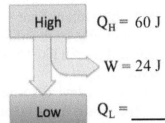

On this page, we'll think about something like a car engine, which uses heat to do work. This is characterized by an energy-flow diagram like the one at right. **Fill in the missing number.**

In a car engine, some energy (Q_H) is extracted by burning gasoline at high temperature (T_H). Only some of this energy can go to do useful work (W), while the rest (Q_L) has to be dumped into a low-temperature reservoir (T_L), such as the environment. Two laws of physics have to be satisfied:

(1) The first law (energy conservation): $Q_H = Q_L + W$ (all absolute values)
(2) The second law: In a nutshell, the second law puts a lower limit on the fraction of heat that just has to be dumped into the low-temperature reservoir. This is the Carnot relationship,

$$\frac{Q_L}{Q_H} = \frac{T_L}{T_H} \qquad \text{and the engine's efficiency is} \qquad e = \frac{W}{Q_H} = 1 - \frac{Q_L}{Q_H}$$

In an ideal case, the Carnot relationship gives Q_L, and in most real situations Q_L is larger than this, but **Q_L cannot be less than this.**

For the nine scenarios shown below, first put a slash through each one that does not satisfy the first law. Then, for the remaining cases, put an X through any that do not satisfy the second law, assuming the engine operates between temperatures of 900K and 300K.

For the remaining cases, write the efficiency below the "Low" box.

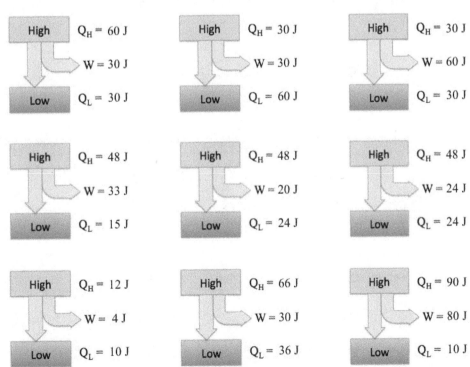

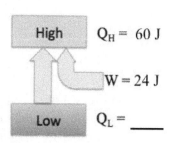

High $\quad Q_H = 60$ J

W = 24 J

Low $\quad Q_L = \underline{\qquad}$

Now, we'll think about something like a refrigerator, air conditioner, or heat pump, which uses work to make heat flow, essentially, from cold to hot. This is characterized by an energy-flow diagram like the one at right. **Fill in the missing number.**

In such a device, some energy (Q_L) is extracted from a region of low temperature (T_L). This requires work (W). The total of the extracted heat and the work (it adds up to Q_H) is then transferred to a high-temperature region (T_H). The same two laws of physics have to be satisfied:

(1) The first law (energy conservation): $Q_H = Q_L + W$ (all absolute values)
(2) The second law: In a nutshell, the second law puts an upper limit on what Q_L can be, for a given Q_H. This is the Carnot relationship,

$$\frac{Q_L}{Q_H} = \frac{T_L}{T_H}$$

The device's coefficient of performance is $\quad COP = \dfrac{Q_H}{W} = \dfrac{1}{1 - \dfrac{Q_L}{Q_H}}$

In an ideal case, the Carnot relationship gives Q_L. In most real situations Q_L is smaller than this, but **Q_L cannot be more than this**. Note: the COP is shown for a heating case, like a heat pump – for a cooling situation, the COP is defined as Q_L / W.

For the nine scenarios shown below, first put a slash through each one that does not satisfy the first law. Then, for the remaining cases, put an X through any that do not satisfy the second law, assuming the device operates between temperatures of 400K and 300K. For the remaining cases, write the COP below the "Low" box.

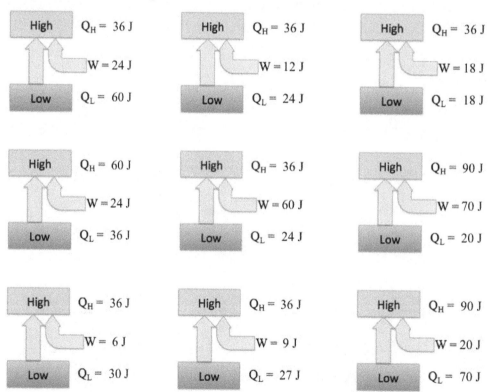

Refrigerators, heat pumps, and air conditioners operate on similar principles to engines, except in these cases work W is added to extract a certain amount of heat Q_L from a low temperature place and deposit an amount $Q_H = W + Q_L$ in a higher-temperature place.

Let's do an example for a heat pump, extracting energy from the cold outdoors (260 K) and depositing it in the warm indoors (300 K). Assuming the heat pump is ideal, and that W = 1000 J, find the values of Q_H and Q_L.

Compare your calculation with the other members of your group.

Write out Newton's universal law of gravitation, and define the variables.

The equation above involves the constant G. What is the name for G and its value?

Show where g = 9.8 m/s/s comes from, at the surface of the Earth.

Write out the general equation for gravitational potential energy.

For this equation, the zero for gravitational potential energy is

[ ] wherever you decide to make it   [ ] at $r = 0$ [ ] at r = infinity

For a small change in height near the surface of the planet, this equation gives the same **change** in potential energy as _____ , the equation we used earlier.

Show how energy conservation is used to find the escape speed of an object from a planet (how fast the object has to be going so it moves away from the planet and never come back).

Gravitational force
Let's move beyond the special-case relationship $\vec{F}_g = m\vec{g}$.

Learning goals – by the end of this section you should be able to
- Apply Newton's law of universal gravitation to understand objects that interact via the force of gravity.
- Use the general equation for gravitational potential energy in applying energy conservation to objects that interact via the force of gravity.

Newton determined that when two objects of mass m and M are separated by a distance r , the magnitude of the gravitational force exerted on one object by the other is given by:

$$F_g = \frac{GmM}{r^2}$$

Newton's Universal Law of Gravitation

The direction is always toward the object applying the force. G is known as the universal gravitational constant, and has a value $G = 6.67 \times 10^{-11}$ N m^2 / kg^2 .

Use the above to express g in terms of the mass M_E of the Earth and the radius R_E of the Earth. Note that $M_E = 5.97 \times 10^{24}$ kg and $R_E = 6.37 \times 10^6$ m .

Using the fact that the gravitational field at the surface of the Earth is about six times larger than that at the surface of the Moon...

and the fact that the Earth's radius is about four times the Moon's radius...

determine how the mass of the Earth compares to the mass of the Moon.

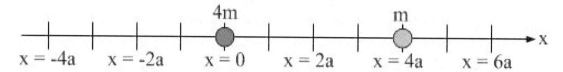

Ball A, with a mass 4m, is placed on the x-axis at x = 0. Ball B, which has a mass m, is placed on the x-axis at x = +4 m. Where would you place ball C, which also has a mass m, so that ball A feels no net force because of the other balls? Is this even possible?

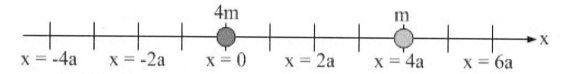

Could you re-position ball C so that ball B is the one feeling no net force? If so, where would you place it?

Kepler's Third Law: See if you can combine the force equation above with our force equation for uniform circular motion, and one more relationship, to derive Kepler's third law, the relationship between the period (T) and radius (r) for a circular orbit:

$$T^2 = \frac{4\pi^2}{GM}r^3$$

Gravitational potential energy
Let's move beyond the special-case relationship U = mgh .

In general, gravitational potential energy is given by the equation $U_g = -\dfrac{GmM}{r}$.

Over relatively small distances, we can use the approximation $\Delta U = mg\,\Delta h$, but for larger distances we need to work with the general equation. Note that the zero of potential energy is defined for us.

Two identical balls, each of mass m, are initially a distance d apart. Each ball is given an initial velocity v, directed away from the other ball. They reach a maximum separation of 3d before reversing direction. The only force acting on each ball once they start is the force of gravity from the other ball. Find v.

Five balls of equal mass, *m*, are placed so that there is one ball at each corner of a regular pentagon. Each ball is a distance *R* from the center. What is the net gravitational field at the center of the pentagon because of these balls?

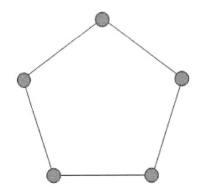

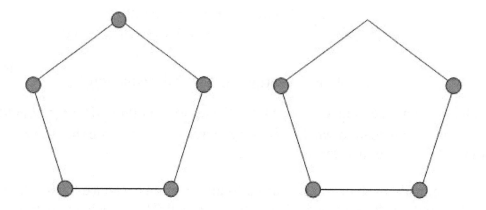

Now, see the picture on the right. The ball at the top vertex has been removed. What is the net gravitational field at the center of the pentagon now because of these balls? (Hint: instead of adding up the contributions from the four remaining balls, try starting from your previous result and subtracting the contribution from the missing ball.)

A ball of mass $6m$ is placed on the x-axis at $x = -2a$. There is a second ball of unknown mass at $x = +a$. The net gravitational field at the origin due to the two balls has a magnitude of Gm/a^2. What is the mass of the second ball? Find all possible solutions.

Additional topics

Relative Velocity

Rolling and Rotational Dynamics

Waves

Relative Velocity
How do we analyze situations involving relative velocity, when one thing is moving with respect to something else?

Consider the following situation, in which Mia and Brandi have a race. The two women are equally fast, but they race with Mia on a moving sidewalk and Brandi on the non-moving floor beside the sidewalk. The women run with constant speed v, while the speed of the moving sidewalk is v/2. They start at a particular spot, run in the same direction as the moving sidewalk to a turn-around point, and then run back to the starting point.

First, predict who wins the race:

Justify your answer conceptually:

Let's say Brandi takes a time T to reach the turn-around point. How long does Brandi's return trip take?

Brandi's time for the entire trip is _____.

In terms of T, how long does it take Mia to reach the turn-around point?

How long does Mia's return trip take?

Mia's time for the return trip is _____.

CHALLENGE QUESTION: Let's say v = 8.0 m/s. If they race a distance L out and L back, where L is 40 m, find all times when the two women are the same distance as one another from the start line.

You're traveling on a train that is heading north at 70 km/h. The conductor is coming to check your ticket - she is walking down the aisle at 5 km/h, heading toward the back of the train. Meanwhile, as you sit in your seat you notice a bicyclist riding his bike on a road that is parallel to the train tracks. Doing some quick calculations, you determine that the bicyclist is moving at 100 km/h relative to you. Fill in the following table to find the relative velocity of all the people/objects with respect to all the other people/objects. The notation $\vec{v}_{YG}$ means "the velocity of You with respect to the Ground."

	You	Ground	Conductor	Bicyclist
You	$\vec{v}_{YY} =$	$\vec{v}_{YG} =$	$\vec{v}_{YC} =$	$\vec{v}_{YB} =$
Ground	$\vec{v}_{GY} =$	$\vec{v}_{GG} =$	$\vec{v}_{GC} =$	$\vec{v}_{GB} =$
Conductor	$\vec{v}_{CY} =$	$\vec{v}_{CG} =$	$\vec{v}_{CC} =$	$\vec{v}_{CB} =$
Bicyclist	$\vec{v}_{BY} =$	$\vec{v}_{BG} =$	$\vec{v}_{BC} =$	$\vec{v}_{BB} =$

Here A, B, and C denote there arbitrary objects. How does $\vec{v}_{AB}$ compare to $\vec{v}_{BA}$?

How do you find $\vec{v}_{AC}$ if you know $\vec{v}_{AB}$ and $\vec{v}_{BC}$? $\vec{v}_{AC} =$

Check your statements against the table above. Note that these expressions work in two dimensions and three dimensions, not just in one dimension.

A particular river is 100 m wide. You have a boat that travels at a constant speed of 5.0 m/s relative to the water. The current in the river is 3.0 m/s downstream.

Sketch a vector diagram, to scale, showing how you would aim your boat so you cross to the far side of the river in the shortest time. If possible, find how long it takes to cross the river and how far upstream or downstream you are when you land on the other side.

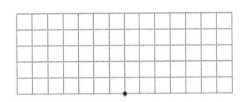

Sketch a vector diagram, to scale, showing how you would aim your boat to land at the point directly across from where you started. How long does it take to cross the river?

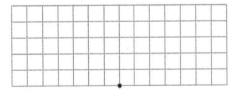

CHALLENGE QUESTION: You now aim your boat not exactly straight across the river but 10° upstream. Determine how long it takes you to cross the river and how far upstream or downstream you are when you land on the other side.

Rolling without Slipping

A cylinder of mass M and radius R has a string wrapped around it, with the string coming off the cylinder above the cylinder. The string is pulled to the right with a force of magnitude F. Our goal is to determine the acceleration of the cylinder if the cylinder rolls without slipping along the horizontal surface.

Sketch this situation and draw a free-body diagram for the cylinder, showing all the forces acting on it.

Step 1 - Apply Newton's Second Law, $\sum \vec{F} = m\vec{a}$.

Step 2 - Apply Newton's Second Law for Rotation, $\sum \vec{\tau} = I\vec{\alpha}$. We usually take torques about the center of the cylinder.

Step 3 – Substitute the rotational inertia of the cylinder, $I = \frac{1}{2}MR^2$, into your torque equation.

Step 4 – For rolling with slipping we can use the relationship $\alpha = \frac{a}{R}$. Substitute that into your equation from step 3.

Step 5 – Combine your equations to find an expression for the cylinder's acceleration, and to find the magnitude and direction of the force of static friction acting on the cylinder.

Angular Momentum
Let's analyze a rotational collision.

Let's analyze a rotational collision, following the basic procedure we followed to analyze a one dimensional collision. Sarah, with mass m and velocity $\bar{v}$, runs toward a playground merry-go-round, which is initially at rest, and jumps on at its edge. Sarah and the merry-go-round (which has mass M, radius R, and $I = cMR^2$) then spin together with a constant angular velocity ω_f. If Sarah's initial velocity is tangent to the circular merry-go-round, what is ω_f?

Is linear momentum conserved here? Why or why not?

Is angular momentum conserved here? Why or why not?

One issue we have in applying angular-momentum conservation is that it's not obvious that there is any angular momentum before the collision, and there clearly is some after the collision. However, a linear momentum can be transformed into an angular momentum in the same way that a force is transformed into a torque.

If $\tau = rF\sin\theta$ then the expression for the angular momentum is $L =$

Write an equation representing angular momentum conservation in this case. Solve it for ω_f.

Rolling and Rotational Kinetic Energy
Let's look at how to incorporate rotational kinetic energy into our energy analysis.

The equation for rotational kinetic energy is analogous to the $\frac{1}{2}mv^2$ equation we're used to for translational kinetic energy. Rotational kinetic energy is $K_{rot} = \frac{1}{2}I\omega^2$. This can be directly incorporated into our usual five-term energy equation.

Take a round object (ball, cylinder, disk, ring, etc.) and roll it down an incline. The object starts from rest at a point that is h higher than the bottom of the incline. Assuming the object rolls without slipping down the incline determine its speed at the bottom.

Sketch this situation, showing the object in two positions, one at the top of the incline and one at the bottom.

Start with the usual conservation of energy equation: $K_i + U_i + W_{nc} = K_f + U_f$. Cross out all the terms that are zero in this equation.

Write out expressions for the remaining terms. Remember to account for both translational kinetic energy and rotational kinetic energy, if appropriate. Keep everything in terms of variables. We'll use as general an analysis as possible, so use $I = cMR^2$, where c will be determined by the shape of the object we use.

Find an expression for the speed of the object at the bottom of the incline. Note that because the object rolls without slipping we can use $\omega = v/R$.

Two objects of equal mass and radius are rolling along a flat surface when they encounter a gradual incline of constant slope. One object is a uniform solid sphere, and the other is a ring. Both objects roll without slipping at all times. Briefly justify all your answers below.

If the objects have identical velocities at the bottom of the incline, which travels farthest up the slope before rolling back down?

[ ] the sphere   [ ] the ring   [ ] they travel the same distance

If the objects instead have identical total kinetic energies at the bottom of the incline, which travels furthest up the slope before rolling back down? The total kinetic energy is the sum of the translational and rotational kinetic energies.

[ ] the sphere   [ ] the ring   [ ] they travel the same distance

Sketch the free-body diagram of one of these objects as it rolls without slipping up the slope. If you include a force of friction clearly indicate whether it is a kinetic force of friction or a static force of friction.

If the objects have identical velocities at the bottom of the incline, which object experiences a larger frictional force as it rolls without slipping up the incline?

[ ] the sphere   [ ] the ring
[ ] neither, they have equal (non-zero) frictional forces
[ ] neither one of the objects has a frictional force

Waves

What is a wave?

How do we classify waves? How many different kinds of waves do you know of?

Let's say we broadcast a wave, somehow, at a single frequency. Do you see any connection between the resulting wave and simple harmonic motion? What?

Let's say I showed you two things.
A – a photograph showing a wave on a string at a particular instant in time.
B – a graph showing the position as a function of time of a single point in the medium.

Which of these two would you use to find the following pieces of information?

The wavelength of the wave?

[ ] A only   [ ] B only   [ ] both A and B

The period of the wave?

[ ] A only   [ ] B only   [ ] both A and B

The maximum speed of a single point in the medium?

[ ] A only   [ ] B only   [ ] both A and B

The speed of the wave?

[ ] A only   [ ] B only   [ ] both A and B

Which of the following determines the wave speed of a wave on a string? Select all that apply.

[ ] the frequency at which the end of the string is shaken up and down

[ ] the coupling between neighboring parts of the string, as measured by the tension in the string

[ ] the mass of each little piece of string, as characterized by μ, the mass per unit length of the string.

The general equation given by some textbooks that describes a transverse wave is:

$y = A\sin\left(2\pi \, ft \pm \dfrac{2\pi x}{\lambda}\right)$, where the + sign indicates that the wave is moving in the $-x$ direction, and the $-$ sign indicates that that wave is moving in the $+x$ direction. We can make the equation more compact by writing it as:

$y = A\sin\left(\omega t \pm kx\right)$.

Here we're using the angular frequency $\omega = 2\pi f = \dfrac{2\pi}{T}$ and the wave number $k = \dfrac{2\pi}{\lambda}$.

Let's use the equation in the following numerical example, where the wave equation describing a particular transverse wave is: $y(x,t) = (0.9 \text{ cm}) \sin[(5.0 \text{ s}^{-1})t - (1.2 \text{ m}^{-1})x]$

Determine the wave's amplitude, wavelength, and frequency.

Determine the speed of the wave.

If the string has a mass/unit length of $\mu = 0.012$ kg/m, determine the tension in the string.

Determine the direction of propagation of the wave. [ ] +x   [ ] –x

Determine the maximum transverse speed of the string.

The general equation given by some textbooks that describes a transverse wave is:

$y = A\sin\left(2\pi ft \pm \dfrac{2\pi x}{\lambda}\right)$, where the + sign indicates that the wave is moving in the –x

direction, and the – sign indicates that that wave is moving in the +x direction. We can make the equation more compact by writing it as:

$y = A\sin\left(\omega t \pm kx\right).$

Here we're using the angular frequency $\omega = 2\pi f = \dfrac{2\pi}{T}$ and the wave number $k = \dfrac{2\pi}{\lambda}$.

Let's use the equation in the following numerical example, where the wave equation describing a particular transverse wave is: $y(x,t) = (0.9\text{ cm}) \sin[(5.0\text{ s}^{-1})t - (1.2\text{ m}^{-1})x]$

Determine the wave's amplitude, wavelength, and frequency.

Determine the speed of the wave.

If the string has a mass/unit length of $\mu = 0.012$ kg/m, determine the tension in the string.

Determine the direction of propagation of the wave. [ ] +x   [ ] –x

Determine the maximum transverse speed of the string.

The Doppler Effect
We'll look at the shift in frequency caused by motion of a source of sound or an observer of that same sound

Is the Doppler effect simply a relative velocity effect? For instance, does a source of sound moving toward you at a particular speed produce the same effect as you moving toward the source of sound at the same speed?

Let's start with the observer moving. Imagine a stationary source broadcasting waves in all directions, and you (the observer) moving through this pattern of waves. How does your motion affect the waves? Effectively, your motion gives the waves a different:

[ ] velocity   [ ] wavelength

Now you, the observer, is stationary, and the source is moving. How does the motion of the source affect the waves? Effectively, the source's motion gives the waves a different:

[ ] velocity   [ ] wavelength

Our general Doppler effect equation gives the observed frequency, f_O, in terms of the frequency emitted by the source, f:

$$f_O = f \left(\frac{v \pm v_O}{v \mp v_S} \right),$$

where v is the speed of sound, v_O is the speed of the observer, and v_S is the speed of the source. In both cases you use the top sign when the motion of the relevant object is toward the other object, and the bottom sign if the motion is away.

A source sends out a frequency of 400 Hz when it is at rest. What is the observed frequency if:
(i) the observer moves toward the source at the speed of sound, and the source is stationary?

(ii) the source moves toward the observer at the speed of sound, and the observer is stationary?

Superposition and Interference of Waves
We'll start looking at what happens when two (or more) waves come together

The principle of superposition: when two waves meet one another, the net displacement of the medium at a particular point equals the sum of the displacements of the waves at that point.

Define **constructive interference**.

Define **destructive interference**.

One example of interference is the phenomenon of beats, which some people use when tuning a guitar, for instance. The basic idea is that you have two sources that have different frequencies, but the frequencies are not different by much. Let's say the sources are initially sending out waves that are in phase, so they interfere constructively. Because the frequencies are different the sources gradually drift out of phase, until eventually we get completely destructive interference. The sources continue to drift out of phase, but this actually moves us back towards constructive interference. The cycle continues.

The beat frequency (the frequency of the louder-softer-louder cycles) equals the frequency difference between the sources.

Example: two tuning forks are labeled 440 Hz. We put an extra mass on the tines of one tuning fork to shift its frequency. When we excite both tuning forks we observe a beat frequency of 5 Hz. What is the frequency of the sound being emitted by the tuning fork with the extra mass?

[ ] 435 Hz   [ ] 440 Hz   [ ] 445 Hz   [ ] 435 Hz or 445 Hz, we can't say for sure

Justify your answer

Consider two pulses propagating in opposite directions along a string. The picture shows the profile of the string at t = 0 and at t = 2 s. Draw the profile of the string at t = 6 s and at t = 8 s.

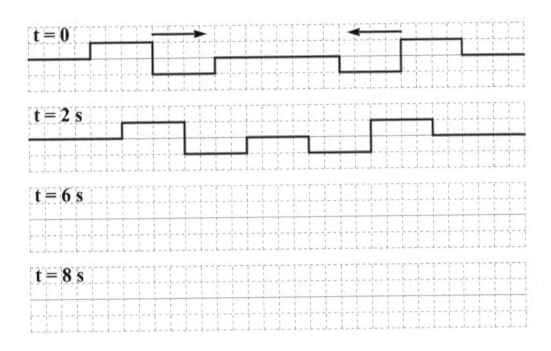

This is the same situation as above, except that the pulse traveling to the left has been inverted. The picture shows the situation at t = 0 s. Draw the profile of the string at
t = 5 s, t = 6 s, and at t = 8 s.

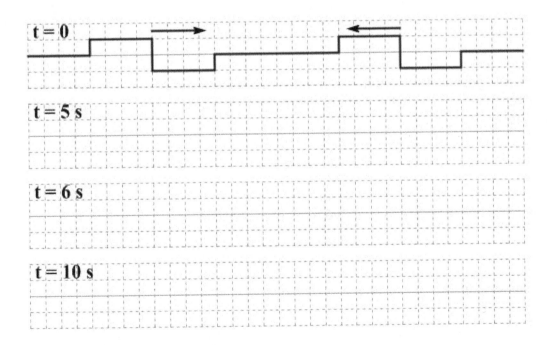

The Interference of Waves
You can get some interesting things when two or more waves come together

Another example of interference comes from two (or more) sources that are broadcasting identical waves. An example is two speakers sending out identical single-frequency sound waves. These sources are separated by some distance, and the waves they broadcast will interfere in the region of space around them; at some points the waves will interfere constructively while at others they will interfere destructively. You might think there would be no logical pattern in this interference, but there actually is a pattern that we can make sense of.

First: pick a point on the perpendicular bisector of the line that joins the two sources. What kind of interference do we expect at any point on the perpendicular bisector? Why?

Next: Move a little bit off the perpendicular bisector, moving in a direction parallel to the line joining the two sources. What happens?

Keep going until you find a point at which the interference is completely destructive. Why? What is it about this point that makes completely destructive interference happen?

Keep going again, until you find a point at which the interference is completely constructive again. What is it about this point that makes completely constructive interference happen?

In general, when does constructive interference occur?

When does destructive interference occur?

Another example of interference is a standing wave. A standing wave can be produced if two identical waves are traveling in opposite directions, and if their wavelength happens to be related to the length of the medium in a particular way.

A common way to create a standing wave (this is how musical instruments work) is to send a wave in one direction along or string (or air column) and let it reflect off the far end. This produces interference between the wave going one way along the string and the wave coming back. We have to understand what happens when a wave reflects off an end, though.

Describe what happens when a wave reflects off a fixed end. An example is the end of a guitar string.

Describe what happens when a wave reflects off a free end. An example is the end of an open pipe.

First let's do an example of interference in two dimensions.

Two sources are broadcasting identical single-frequency waves, in phase. You stand 3 m from one source and 4 m from the other. What is the lowest frequency at which constructive interference occurs at your location? Take the speed of sound to be 340 m/s.

What are the next two smallest frequencies at which constructive interference occurs?

What are the lowest three frequencies giving destructive interference at your location?

Standing Waves
Understanding standing waves can help us understand musical instruments

A standing wave is produced when two identical waves are traveling in opposite directions. For musical instruments, we get constructive interference for the waves if their wavelength happens to be related to the length of the medium in a particular way.

A common way to create a standing wave (this is how musical instruments work) is to send a wave in one direction along or string (or air column) and let it reflect off the far end. This produces interference between the wave going one way along the string and the wave coming back. We have to understand what happens when a wave reflects off an end, though.

Describe what happens when a wave reflects off a fixed end. An example is the end of a guitar string.

Describe what happens when a wave reflects off a free end. An example is the end of an open pipe.

Sketch the three lowest-frequency standing waves for a string fixed at both ends.

$$f_n = \frac{nv}{2L} \quad (n = 1, 2, 3, ...)$$

Lowest (the fundamental):

Next lowest (the second harmonic):

Next lowest (the third harmonic):

Sketch the three lowest-frequency standing waves for a tube open at one end and closed at the other. $f_n = \dfrac{nv}{4L}$ (n = 1, 3, 5, ...)

Lowest (the fundamental):

Next lowest (the third harmonic):

Next lowest (the fifth harmonic):

For a pipe that has only one end closed, it is found that the frequencies of two successive harmonics differ by 100 Hz. What is the pipe's fundamental frequency?

What is the length of the pipe?

CPSIA information can be obtained
at www.ICGtesting.com
Printed in the USA
LVHW021618140719
624039LV00011B/456